Human–Computer Interaction

Fundamentals and Practice

Human–Computer Interaction

Fundamentals and Practice

Gerard Jounghyun Kim

CRC Press is an imprint of the
Taylor & Francis Group, an **Informa** business

CRC Press
Taylor & Francis Group
6000 Broken Sound Parkway NW, Suite 300
Boca Raton, FL 33487-2742

Printed on acid-free paper
Version Date: 20141208

International Standard Book Number-13: 978-1-4822-3389-6 (Hardback)

the Taylor & Francis Web site at
www.taylorandfrancis.com

CRC Press Web site at
w.crcpress.com

Contents

Preface

Human–computer interaction (HCI) is becoming ever more important in interactive software. Such software has long been evaluated in terms of the availability and breadth of its functions and its algorithmic efficiency. While such a developer's perspective is still somewhat valid, it has become difficult to differentiate among similar software components from such an aspect given the amazing computing performance of today's hardware and the spread of algorithmic knowledge and systems development know-how. Thus software quality is increasingly judged from the users' external point of view in terms of their expectations, satisfaction, and experience. This external view or user experience may be defined in many ways, but it is most obvious that it has quite a lot to do with how the software users interact with it and, hence, its design. HCI will become even more critical as everything around us becomes digital and unknowingly embedded with interactive computing services that make our everyday lives more exciting, efficient, and convenient.

Therefore, software (at least software that is highly interactive and targeted for a high number of users) must now be developed with HCI as one of its higher priorities. However, at the undergraduate level, it is still often the case that HCI is not given the attention it deserves in the education of future software developers. Most entry-level HCI textbooks are structured around high-level concepts and guidelines and are not directly tied to the software development process. Some of these books may offer design patterns, but students at the undergraduate level might still find it puzzling as to how HCI fits in with their basic software development knowledge. In fact, most of the HCI concepts and guidelines are fairly commonsense or very easy to comprehend. (After all, how difficult would it be to make one understand that users are important?) But it is in the practice and within the context of actual development that one has to make the difficult choices to produce highly usable interactive software.

Following this line of thinking, this book was designed around the overall development cycle for an interactive software product. It starts with the required basic HCI knowledge, which is kept as compact as possible by including only the basic essentials (Chapters 1–3). The intention is to convey the spirit of HCI rather than a long list of compiled knowledge. The book then moves into the application of this knowledge by iteratively forming the HCI requirements and modeling the interaction process (Chapter 4), designing the interface (Chapter 4), implementing the resulting design (Chapters 5–7), and finally evaluating the implemented product (Chapter 8). The book is targeted mainly at undergraduate students of computer science and information technology (IT), but it is easy enough to be taken up by readers in other fields. Some knowledge of computers and programming would be desirable, but it is not absolutely necessary. (Those not interested in the detailed aspects of implementation can skip some of Chapters 5–7.)

The core content of the book is based on the introductory undergraduate HCI course (advanced junior or senior level) that I have taught since 2006 at Korea University. The following table shows how one might structure a similar course using this book (or pace oneself for self-teaching).

	Lecture
Weeks 1–2	Chapters 1–2: Introduction, HCI principles, and guidelines
Weeks 3–5	Chapter 3: Cognitive science, GOMS, human factors
	Homework 1: • Application of HCI principles/guidelines • GOMS exercise
Weeks 6–8	Chapter 4: HCI design

	Homework 2: • Project proposal (Part 1): Functional and UI requirements, user analysis, etc. • Design of the app (Part 2): Interaction model, scenario, storyboards, basic interface design, and wire-framing • Short presentation
Week 9	Midterm exam (Chapters 1–4)
Weeks 10–11	Chapters 5–7: Implementation issues Homework 3: • First implementation of project (using the MVC model) • Presentation (MVC structure) and working demo 1
Weeks 12–13	Chapter 8: Evaluation
Weeks 14–15	Chapter 9: Future of HCI Homework 4: • Self-heuristic evaluation for the project • Carry out and receive peer review for other projects and one's own project • Redesigning/reimplementation of the project app • Presentation of "before" and "after" and working demo 2
Week 15/16	Final exam

The PowerPoint lecture slides and the source code for the example application used in this book ("No Sheets 1.0," also downloadable through Google Play) are available through the publisher's resource website (see http://www.crcpress.com/product/isbn/9781482233896). I sincerely hope that the book will help readers to develop and acquire an HCI mindset as an important step to becoming a capable IT professional in the field.

The completion of this book was possible only with the greatest help and understanding from many people. My first thanks go to my graduate students at the Digital Experience Laboratory at Korea University (Youngsun, Youngwon, Changhyun, Jong-gil, Sang-yong, Jae-dong, Myong-hee, and Euijae). They helped me with proofreading, drawing figures, formatting, and many other tasks in the midst of research, projects, classes, and all the other things that make up the life of a graduate student. My dear colleagues in the HCI community have also given me much valuable feedback regarding the content and structure of the book. In particular, I thank Prof. Jee-in Kim, Dr. Gun Lee, Prof. Woontak Woo, Prof. Jinwoo Kim, Prof. Jongwon Lee, Prof. Jong-il Park, Prof. Seokhee Jeon, Prof. Si-Jung Kim, Dr. Ungyeon Yang, Prof. Junho Kim, Prof. Chang-Guen Song, Prof. Jin-seok Seo, Prof. Sookjin Kim, Prof. Junho Choi, and Prof. Mincheol Hwang. I am very grateful for the support of the

KRF-funded Engineering Center of Kwangwoon University (head: Professor Eunsoo Kim). CRC Press has been very patient and prompt with assistance for all my writing problems, not to mention seeing the value in publishing this book. Finally, I thank my wife Sooah, my parents, and my children (Andrew and Ellen) for their understanding and just for being there!

About the Author

Gerard "Gerry" Jounghyun Kim earned his bachelor's in electrical and computer engineering at Carnegie Mellon University and his master's and PhD at the University of Southern California. He started his academic career at POSTECH in 1996 after a short post at the U.S. National Institute of Standards and Technology as a National Research Council postdoctoral fellow. In 2006, he moved to Korea University. Since 1996, he has conducted research in the field of HCI, including virtual and mixed reality, mobile interaction, and multimodal interaction. Dr. Kim has written more than 100 articles in international and domestic journals and conferences, and he is the author of *Designing Virtual Reality Systems* (Springer, 2005).

1
INTRODUCTION

1.1 What HCI Is and Why It Is Important

Human–computer interaction (HCI) is a cross-disciplinary area (e.g., engineering, psychology, ergonomics, design) that deals with the theory, design, implementation, and evaluation of the ways that humans use and interact with computing devices. *Interaction* is a concept to be distinguished from another similar term, *interface.* Roughly speaking, interaction refers to an abstract model by which humans interact with the computing device for a given task, and an interface is a choice of technical realization (hardware or software) of such a given interaction model. Thus, the letter *I* in HCI refers to both interaction and interface, encompassing the abstract model and the technological methodology (Figure 1.1).

HCI has become much more important in recent years as computers (and embedded devices) have become commonplace in almost all facets of our lives. Aside from merely making the necessary computational functionalities available, the early focus of HCI has been in how to design interaction and implement interfaces for high usability. The term *high usability* means that the resulting interfaces are easy to

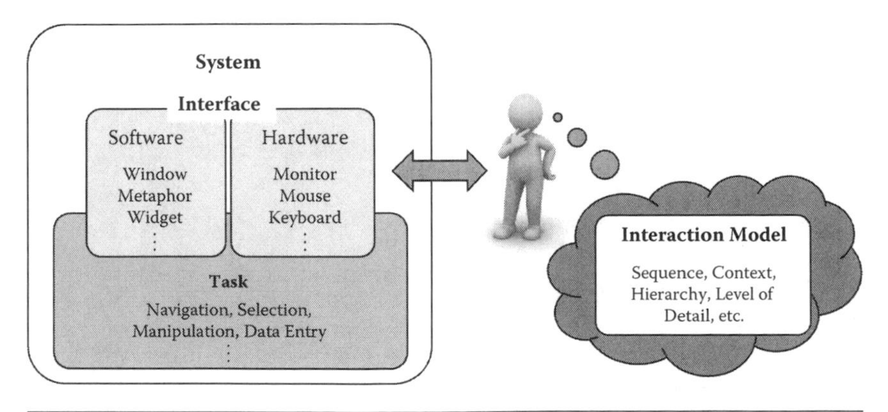

Figure 1.1 The distinguishing concepts of interaction (model) and interface.

use, efficient for the task, ensure safety, and lead to a correct completion of the task. Usable and efficient interaction with the computing device in turn translates to higher productivity.

The simple aesthetic appeal of interfaces (while satisfying the need for usability) is now a critical added requirement for commercial success as well. The family of distinctly designed Apple® products is a good example. Apple products are attractive and have created a multitude of faithful followers even though their functionality may be virtually equal to their competitors. In this context, the concept of *user experience* (UX) has lately become a buzzword, a notion that not only encompasses the functional completeness, high usability, and aesthetic appeal of the interactive artifact, but also its seamless integration into one's lifestyle or even creating a new one around it (Figure 1.2).

A less acknowledged fact is how HCI has had a huge impact in the history of computing and changed our daily lives. It was probably the invention (or rediscovery) of the mouse that was the linchpin in the personal

Figure 1.2 Goals of human–computer interaction (HCI): (a) functional completeness (Apple iPhone 5s, http://www.apple.com/iphone-5s), (b) high usability (Microsoft® Pixelsense, http://blogs.msdn.com/b/pixelsense), (c) aesthetic appeal (Apple iPhone 5s), and (d) compelling user experience (UX) (Microsoft Kinect, http://www.xbox.com/ko-KR/Kinect).

Figure 1.3 The evolution of interfaces in the course of the history of computing (i.e., terminal and keyboard, graphic user interface and mouse, and handheld and touch-based interface). (Courtesy of Cox, J., https://www.flickr.com/photos/15587432@N02/3281139507, Melbourne, FL.)

computer revolution, making the operation of a computer intuitive and much easier than the previous system of keyboard commands. The spreadsheet interface made business computing a huge success. The Internet phenomenon could not have happened without the web-browser interface. Smartphones, with their touch-oriented interfaces, have nearly replaced the previous generation of feature phones. Body-based and action-oriented interfaces are now introducing new ways to play and enjoy computer games. HCI still continues to redefine how we view, absorb, exchange, create, and manipulate information to our advantage (Figure 1.3).

1.2 Principles of HCI

Despite its importance, good HCI design is generally difficult, mainly because it is a multiobjective task that involves simultaneous consideration of many things, such as the types of users, characteristics of the tasks, capabilities and cost of the devices, lack of objective or exact quantitative evaluation measures, and changing technologies, to name just a few. A considerable knowledge in many different fields is required. Over the relatively young history of HCI, researchers and developers in the field have accumulated and established basic principles for good HCI design in hopes of achieving some of the main objectives (as a whole) that were laid out in the previous section. These HCI principles are general, fundamental, and commonsensical, applicable to almost any HCI design situation. Here, we provide a short review of the main HCI principles.

1.2.1 "Know Thy User"

The foremost creed in HCI is to devise interaction and interfaces around the target users. This overall concept was well captured by the phrase,

"Know thy user," coined by Hansen [1] in 1971, even though the so-called user-centered design approach has become a buzzword only in recent years. This principle simply states that the interaction and interface should cater to the needs and capabilities of the target user of the system in design. However, as easy as this sounds, it is more often the case that the HCI designers and implementers proceed without a full understanding of the user, for example, by just guessing and pretending to know and be able to predict how the representative user might respond to one's design. Ideally, comprehensive information (e.g., age, gender, education level, social status, computing experience, cultural background) about the representative target user should be collected and analyzed to determine their probable preferences, tendencies, capabilities (physical and mental), and skill levels. Such information can be used to properly model interaction and pick the right interface solution for the target users.

Consider a situation where a developer is working to change an interface, supposedly to achieve higher usability. However, we might need to remember that while young adults are extremely adept at and open to adopting new interfaces, older generations are much less so. Here is another example. Males are generally known to be better than females in terms of spatial ability and, as such, one might consider such a fact in employing three-dimensional (3-D) user interfaces. However, other studies point to females majoring in engineering and science to possess an equivalent level of spatial ability as their male counterparts [2]. So sometimes, conventional wisdom alone may not be sufficient to warrant proper interface design. These examples illustrate that there are a great many aspects that need to be considered in this regard. If a direct field study is not feasible, an experienced and humble HCI designer will at least try to leverage the vast knowledge available from cognitive psychology, ergonomics, and anthropomorphic data to assess the capabilities and characteristics of the target user group. Figure 1.4 shows examples of user-centered designs of web pages for kids and the elderly.

A related (or perhaps even opposing) notion to the user-centered design is the concept of "universal usability," which roughly promotes "humane" interfaces that cater to a wide (rather than a specific) range of users, i.e., across age groups, skill levels, cultural backgrounds, and disability levels. Such a notion has become almost required in our advanced multicultural societies. However, as wonderful as it sounds, it is generally very difficult to achieve this with a single interface.

Figure 1.4 Examples of user-centered designs of web pages for (a) kids (courtesy of Junior Naver, http://jr.naver.com), and (b) the elderly (courtesy of SilverNet News, http://www.silvernews.or.kr).

Usually, universal usability is achieved by justifying the investment required to build separate interfaces for distinct user groups. For example, in advanced countries, many government web pages are now legally required to provide interfaces in different languages and for color-blind and visually challenged users (Figure 1.5). Many interactive systems provide both menu-driven commands for novices and keyboard-based hot keys for experts (Figure 1.6).

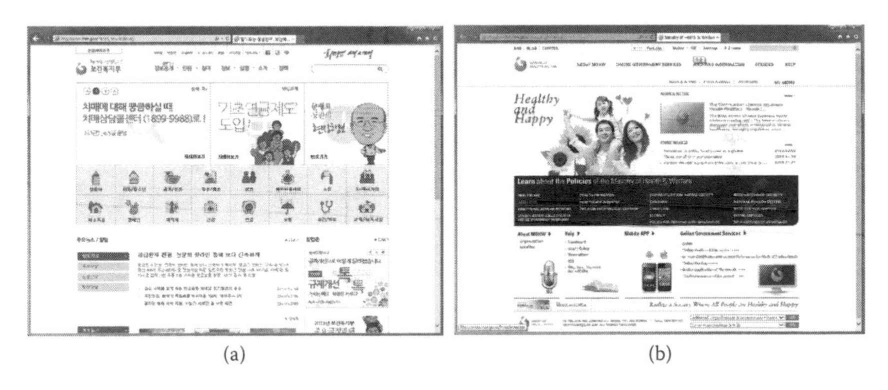

Figure 1.5 Two different interfaces to achieve universal usability (one in Korean and the other in English). (From the Korean Ministry of Health and Welfare, http://english.mw.go.kr/front_eng/index.jsp.)

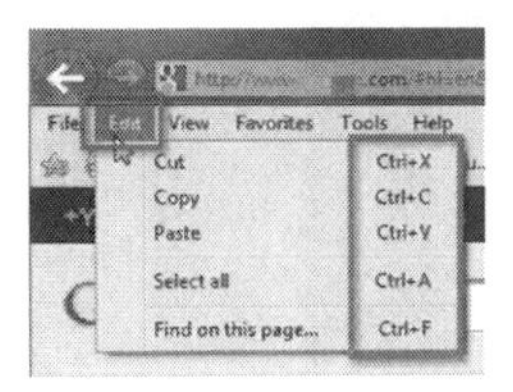

Figure 1.6 An interface providing both menus (for novice users) and hot keys (for expert users).

1.2.2 Understand the Task

Another almost-commonsensical principle is to base HCI design on the understanding of the task. The term *task* refers to the job to be accomplished by the user through the use of the interactive system. In fact, understanding the task at hand is closely related to the interaction modeling and user analysis. It really boils down to identifying the sequence and structure of subtasks at an abstraction level appropriate for the typical user within the larger application context. Take the subtask (for a larger application) for "changing the Wi-Fi connection access point" for a smartphone. For an expert user experienced in computer networks, the task might be modeled with detailed steps, asking the user to select from a pool of available nearby access points based on their characteristics such as the signal strength, bandwidth, security level, and so forth. On the other hand, for a casual user, the subtask might only involve entering a password for the automatically selected access point (Figure 1.7).

Note again that the task (or, equivalently, the interaction) model must ideally come from the user. Different users will have different mental models of the task at hand, and this must be reflected in the structure of the interface to simplify implementation for all users. We will study the process of task/interaction modeling in Chapter 2 in more detail. However, it is not always the case that modeling interaction after the user is the most efficient approach. One must remember that humans are very adaptive and, as such, a nonuser-based

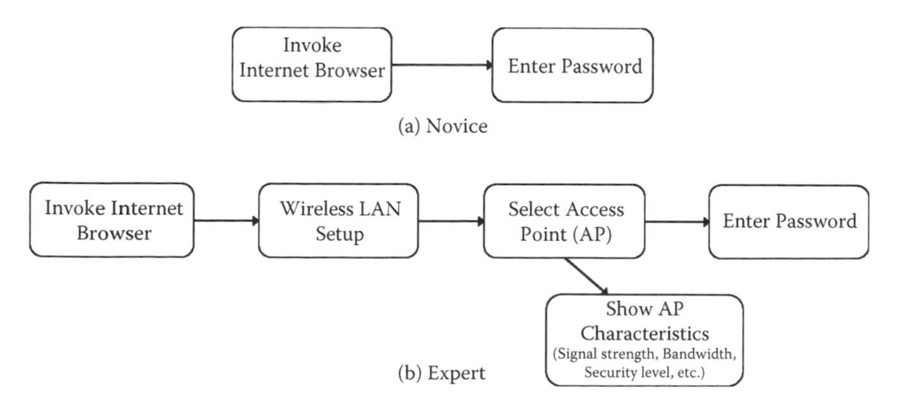

Figure 1.7 Two interaction models at different levels of detail for the task of "connecting to the Internet from a smartphone," depending on the user type.

task/interaction model may sometimes be developed based solely on the general human capacity.

1.2.3 Reduce Memory Load

Designing interaction with as little memory load as possible is a principle that also has a theoretical basis. Humans are certainly more efficient in carrying out tasks that require less memory burden, long or short term. Keeping the user's short-term memory load light is of particular importance with regard to the interface's role as a quick and easy guidance to the completion of the task. The capacity of the human's short-term memory (STM) is about 5–9 chunks of information (or items meaningful with respect to the task), famously known as the "magic number" [3]. Light memory burden also leads to less erroneous behavior. This fact is well applied to interface design, for instance, in keeping the number of menu items or depth to less than this amount to maintain good user awareness of the ongoing task or in providing reminders and status information continuously throughout the interaction (Figure 1.8).

1.2.4 Strive for Consistency

In the longer term, one way to unburden the memory load is to keep consistency [4]. This applies to (a) both within an application and across

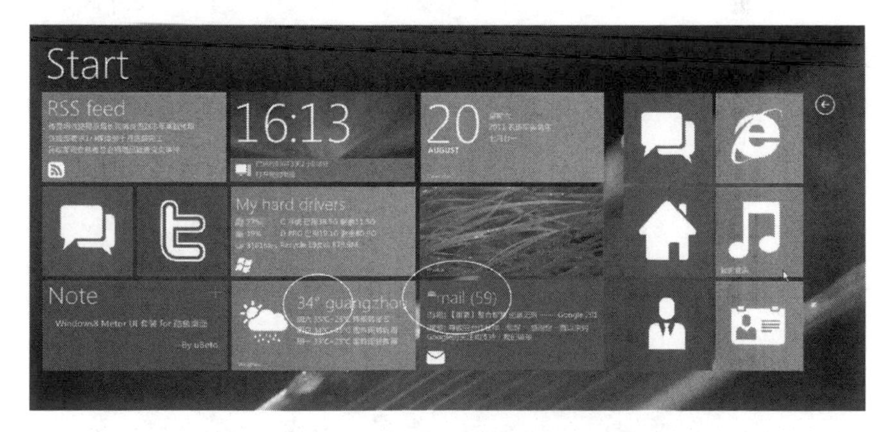

Figure 1.8 Interfaces designed for minimal short-term memory: (a) a menu system with fewer than 10 items (left) and (b) categorization by colors, areas, icons, and labels. Badges are used to display status information such as the current weather (see circled portions) and number of unread mails as a constant reminder. (From Microsoft®, Microsoft Metro interface, http://www.microsoft.com.)

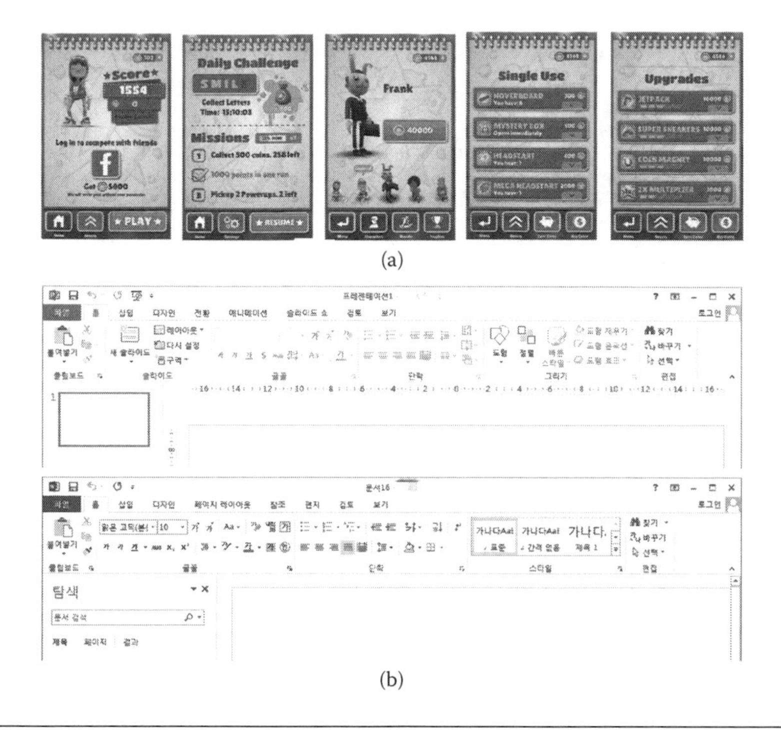

Figure 1.9 (a) A consistent look of the interface within an application (a game called Subway Surfers, https://play.google.com/store/apps/details?id=com.kiloo.subwaysurf) and (b) a consistent interface between Microsoft PowerPoint and Word.

different applications and (b) both the interaction model and interface implementation. For instance, the user is likely to get confused and exhibit erroneous responses if the same subtask is involved, at different times, for different interaction steps or interface methods. Note that the exact same subtasks may appear across different applications as well. Aside from being able to remember what to do, consistency and familiarity also lead to higher acceptability and preference. One way the Microsoft Windows®–based applications maintain their competitiveness is by promoting consistent and familiar interfaces (Figure 1.9).

1.2.5 Remind Users and Refresh Their Memory

Any significant task will involve the use of memory, so another good strategy is to employ interfaces that give continuous reminders of important information and thereby refresh the user's memory. The human memory dissipates information quite quickly, and this is especially true when switching tasks in multitasking situations (which is a very prevalent form of interaction these days). In fact, research shows that our brain internally rehearses information encoding during multitasking [5]. Even a single task may proceed in different contextual spans. For instance, in an online shopping application, one might cycle through the entry of different types of information: item selection, delivery options, address, credit card number, number of items, etc. To maintain the user's awareness of the situation and further elicit correct responses, informative, momentary, or continuous feedback will refresh the user's memory and help the user complete the task easily.

One particular type of informative feedback (aside from the current status) is the reaffirmation of the user action to signal the closure of a larger process [6]. An example might be not only explicitly confirming the safe receipt of a credit card number, but also signaling that the book order is complete (and "closed"). Such a closure will bring satisfaction by matching the user's mental picture of the ongoing interactive process (Figure 1.10).

1.2.6 Prevent Errors/Reversal of Action

While supporting a quick completion of the task is important, error-free operation is equally important [6]. As such, the interaction and interface should be designed to avoid confusion and mental overload. Naturally, all of the aforementioned principles apply here. In addition, one effective technique is to present or solicit only the relevant information/action as required at a given time. Inactive menu items are good examples of such a technique. Also, having the system require the user to choose from possibilities (e.g., menu system) is

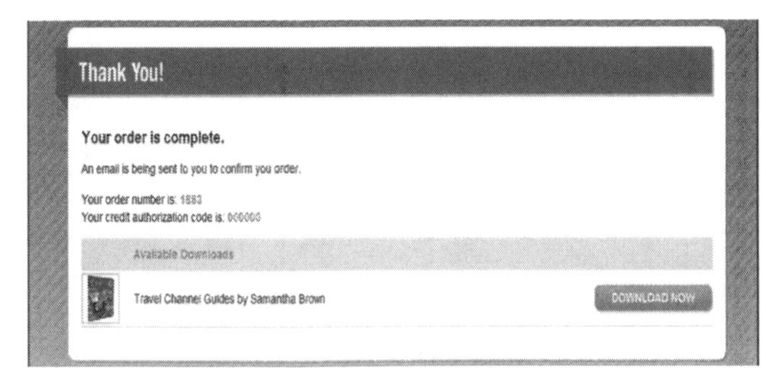

Figure 1.10 Reaffirming the user's action (i.e., credit card number correctly and securely entered) and a larger interactive process (i.e., the book purchase is complete).

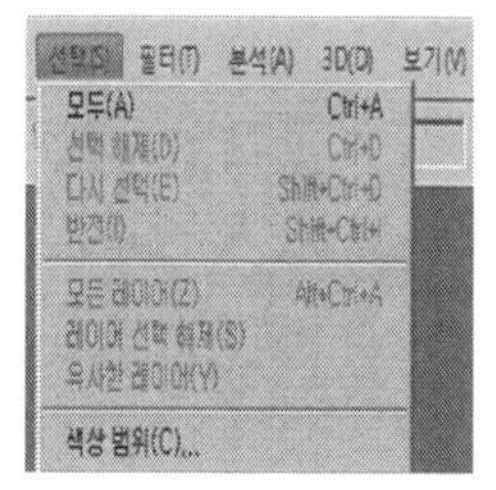

Figure 1.11 Preventing errors by presenting only the relevant information at a given time (inactive menu items) and making selections rather than enforcing recall or full manual input specification.

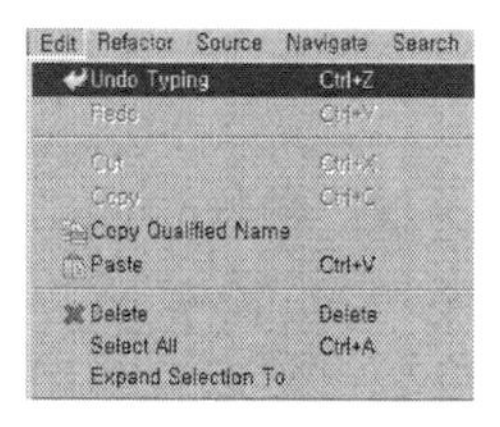

Figure 1.12 Making the user comfortable by always allowing an easy reversal of action.

generally a safer approach than to rely on recall (e.g., direct text input) (Figure 1.11).

Despite employing some of the principles and techniques described here, there is always a chance that the user will make mistakes. Thus, a very obvious but easy-to-forget feature is to allow an easy reversal of action. This puts the user into a comfortable state and increases user satisfaction as well (Figure 1.12).

1.2.7 Naturalness

The final major HCI principle is to favor "natural" interaction and interfaces. *Naturalness* refers to a trait that is reflective of various operations in our everyday life. For instance, a perfect HCI may one day be realized when a natural language–based conversational interface is possible, because this is the prevalent way that humans communicate. However, it can be tricky to directly translate real-life styles and modes of interaction to and for interaction with a computer. Perhaps a better approach is to model interaction "metaphorically" to the real-life counterpart, extracting the conceptual and abstract essence of the task. For instance, Figure 1.13 shows an interface called the ARCBall [7] for rotating an object in 3-D space using a mouse (2-D device). In

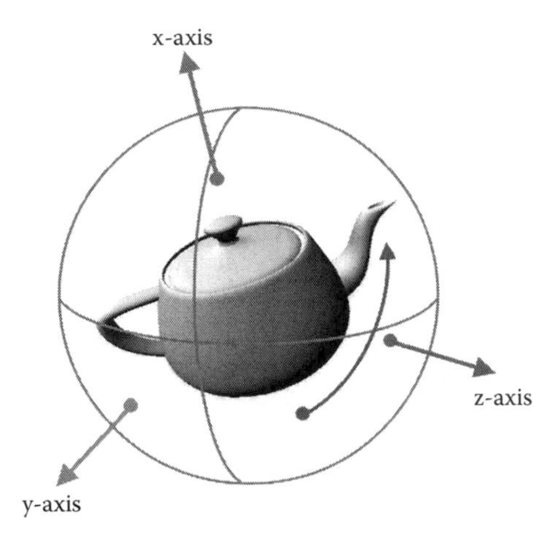

Figure 1.13 ARCBall: 3-D object rotation by using the sphere metaphor. It is also very intuitive with a high level of affordance. (From Shoemake, K., *Graphics Interface*, 92, 151–156, 1992 [7].)

order to rotate, the selected object is overlaid with and enclosed by a transparent sphere, and the user drags on the surface of the sphere to rotate the object inside. One might consider this rotation technique to be metaphoric because it abstracts the interaction object into the shape of a sphere, the most rotational object we know.

A natural or metaphoric interface (assuming that the metaphor is not contrived) will also have *affordance*, a property (or additional cues) that appeals to our innate perception and cognition, thus making it so intuitive that the interface would require almost no learning [4]. In the example of the ARCBall, the spherical shape of the rotator GUI may be regarded to exhibit a high level of affordance, requiring no explanation as to how to rotate the object.

1.3 Summary

In this chapter, I have introduced the field of HCI, namely its objective and importance. We also have reviewed some of the main high-level principles of HCI and presented some relevant examples. These principles are often based on or are just manifestations of deeper theories in cognitive science and ergonomics. However, they are transformed into more detailed and directly usable guidelines when put into actual practice for the specific purpose of designing an effective interface. In Chapters 2 and 3, we take a look at these guidelines and theories, respectively, as they are essential knowledge required for the HCI design process, which we will begin to address in Chapter 4.

References

1. Hansen, Wilfred J. 1971. User engineering principles for interactive systems. In *AFIPS '71 (Fall) Proceedings of the November 16–18, 1971, Fall Joint Computer Conference*, 523–32. New York: ACM.
2. Eisenberg, Theodore A., and Robert L. McGinty. 1977. On spatial visualization in college students. *Journal of Psychology* 95 (1): 99–104.
3. Miller, George A. 1956. The magical number seven, plus or minus two: Some limits on our capacity for processing information. *Psychological Review* 63 (2): 81–97.
4. Norman, Donald A. 2002. *The design of everyday things*. New York: First Basic Paperback. (Orig. pub. 1988 as *Psychology of everyday things*. New York: Basic Books.)
5. Salvucci, Dario D., and Neils A. Taatgen. 2010. *The multitasking mind*. Oxford, UK: Oxford University Press.
6. Shneiderman, Ben, and Catherine Plaisant. 2004. *Designing the user interface: Strategies for effective human–computer interaction*. 4th ed. Boston: Addison Wesley.
7. Shoemake, Ken. 1992. ARCBALL: A user interface for specifying three-dimensional orientation using a mouse. *Graphics Interface* 92:151–56.

2
Specific HCI Guidelines

2.1 Guideline Categories

While principles are very general and applicable to wide areas and aspects of human–computer interaction (HCI) design, guidelines tend to be more specific. Table 2.1 shows major criteria and areas for which specific guidelines can be of help in HCI design. For instance, in the criterion of "user type," there could be further specific guidelines for specific age groups or gender.

Many guidelines in the categories listed in Table 2.1 have been put forth by a number of HCI researchers, practitioners, and organizations over the years and are considered to be reasonably objective. There is even an international standard; the International Organization for Standardization (ISO) 9241 document guides the ergonomics aspects of HCI designs, with topics covering visual display, physical input devices, workplace/environment ergonomics, and tactile/haptic interactions [1]. Broadly, we might divide the guidelines into two categories: (a) domain specific (i.e., specific to user, platform, etc.) and (b) of general HCI design. Note that these guidelines can be relevant and common across the different categories shown in Table 2.1. For example, guidelines for e-commerce application might also address different general HCI design issues such as display layout, how to solicit input, how to promote vendor-specific styles, and how to target for a particular user group.

Even though guidelines are much more specific than the principles, it is still not very clear how to reflect them into the HCI design in a concrete and consistent manner. In this regard, Tidwell has compiled many user interface (UI) design patterns in the form of guidelines [2]. Tidwell's guidelines address many categories of the "general HCI design" issues (see Table 2.1) such as display layout, information

Table 2.1 Examples of Criteria/Categories for HCI Guidelines

CRITERIA	MAIN CATEGORIES	EXAMPLES
User type	Age/generation Disability/accessibility Gender Consumer group Occupation Culture/country	Kids, elders, visually challenged, baby boomers, students, parents, East Asians, athletes, etc.
Platform/system setup	Mobile/handheld Desktop Large display/virtual reality Embedded Public installation Operating system/network	Smartphone, padlike device, desktop, kiosk, embedded OS, cloud based, navigation systems, personal game players, MP3 players, e-book, etc.
Vendors/organizations	Private Public Design style/identity	NASA, Korea University, Android™, iOS, Windows® XP, etc.
Interface style/ modality/technology	WIMP Non-WIMP 3-D Multimodal	Voice/aural, gesture, single/ multitouch, tactile/haptic, multimodal, menu driven, GUI/ widgets, visual perception, etc.
Task/operational context	Location/place Time Noise/lighting Bodily constraints	Office, outdoor, road/street, home, automobile, subway, classroom, eyes free, hands free, handedness, etc.
Applications	Game Media/information Electronic commerce Design/editing Social network service	
General HCI design	Display layout Information structure/navigation Soliciting input Information/output visualization Design process and practices User experience General aesthetics	

[a] WIMP is an acronym for *windows, icon, mouse,* and *pointer*, which represents the conventional desktop interface.

structure and navigation, as well as data entry and even aesthetic aspects. Each guideline illustrates specific UI examples with exact descriptions of what it is and what it does and why and when it should be used. Such design patterns are of great help during actual HCI design.

It is not possible to list and explain all the guidelines that exist for all the various areas. Despite differences in the specifics, most of them are commonly shared and equivalent or can be understood in terms of the higher level principles. Here we present a few examples.

2.2 Examples of HCI Guidelines

2.2.1 Visual Display Layout (General HCI Design)

One of the main focuses in many design guidelines is on the display (page) layout. This problem concerns organizing and allotting relevant information (both the content and UI elements) in one visible screen or scrollable page. Generally, the display layout should be such that it is organized according to the information content (e.g., importance, sequence, functionality), is sized manageably (e.g., divided into proper sections), is attention grabbing, and is visually pleasing (e.g., aligned and with restricted use of colors). Table 2.2 is a summarized guideline for web-page layout put forth by the U.S. Department of Health and Human Services (HHS) for the US government [3].

Table 2.2 Examples of Guidelines for Government Web Page Layout

GUIDELINES	EXPLANATION
Avoid cluttered displays	Create pages that are not considered cluttered by users
Place important items consistently	Put important, clickable items in the same locations and closer to the top of the page, where their location can be better estimated
Place important items at top center	Put the most important items at the top center of the web page to facilitate users finding the information
Structure for easy comparison	Structure pages so that items can be easily compared when users must analyze those items to discern similarities, differences, trends, and relationships
Establish level of importance	Establish a high-to-low level of importance for information and infuse this approach throughout each page on the website
Optimize display density	To facilitate finding target information on a page, create pages that are not too crowded with items of information
Align items on a page	Visually align page elements, either vertically or horizontally
Set appropriate page lengths	Make page-length decisions that support the primary use of the web page
Choose appropriate line lengths	If reading speed is most important, use longer line lengths (75–100 characters per line); if acceptance of the website is most important, use shorter line lengths (50 characters per line)
Use frames when functions must remain accessible	Use frames when certain functions must remain visible on the screen as the user accesses other information on the site

Source: Leavitt, M. O., and Shneiderman, B., *Research-Based Web Design and Usability Guidelines*, U.S. Department of Health and Human Services, Washington, DC, 2006 [3].

2.2.2 Information Structuring and Navigation (General HCI Design)

A single display is often not sufficient to encompass all of the required information content or to control the UI for a given application. Thus, structuring the information and making it easy to move (or navigate) among the various items becomes a very important issue for high usability. Structuring information content and controlling the interface for the purpose of HCI is closely related to the principle of understanding the task (Section 1.2.2). By understanding the task, we identify the sequence of subtasks and actions, and each task will be associated with information either for making input or for the resulting output. The task structure, action sequence, and associated content organization will dictate the interaction flow and its fluidity. In this way, only the right amount of information or control will be available at the right time.

Aside from such internal structure, it is also important to provide external means and the right UI for fast and easy navigation. Fast and easy navigation means enabling the user to find the needed action (e.g., menu item) and information quickly (Figure 2.1). Here, we introduce a summarized guideline for the design of an easily navigated interface from Leavitt and Shneiderman [3].

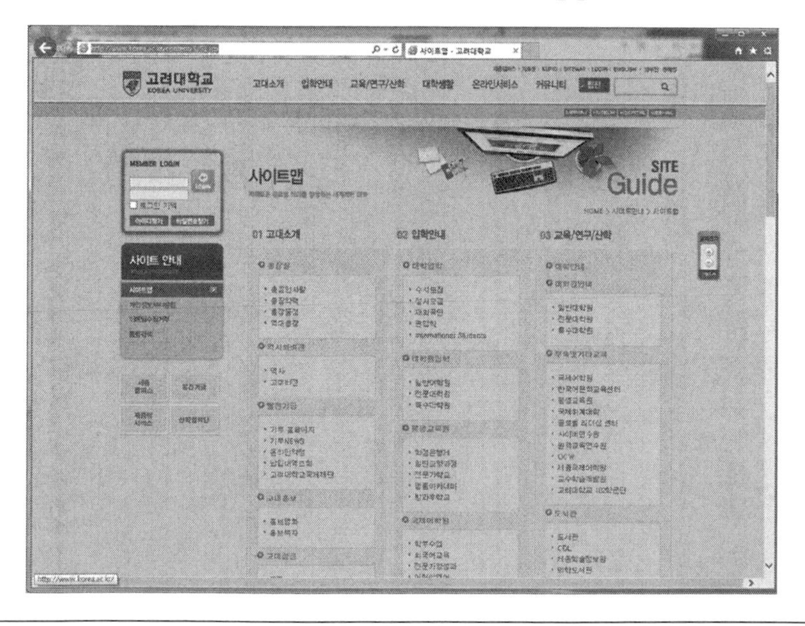

Figure 2.1 An example of a site map for a website. (From Korea University, http://www.korea.ac.kr. With permission.)

Navigation refers to the method used to find information within a Web site. A navigation page is used primarily to help users locate and link to destination pages. A Web site's navigation scheme and features should allow users to find and access information effectively and efficiently. When possible, this means designers should keep navigation-only pages short. Designers should include site maps, and provide effective feedback on the user's location within the site....

...To facilitate navigation, designers should differentiate and group navigation elements and use appropriate menu types. It is also important to use descriptive tab labels, provide a clickable list of page contents on long pages, and add "glosses" where they will help users select the correct link. In well-designed sites, users do not get trapped in dead-end pages.

As a more concrete example, we illustrate two design patterns from Tidwell [2]. Note that as design patterns, very specific uses of UI elements are suggested addressing the concerned issue (Figures 2.2 and 2.3).

What:

Put two side-by-side panels on the interface. In the first, show a set of items that the user can select at will; in the other, show the content of the selected item.

Use when:

You're presenting a list of objects, categories, or even actions.... You want the user to see the overall structure of the list....
Physically, the display you work with is large enough to show two separate panels at once....

Figure 2.2 The use of a two-panel selector, a design pattern for information structuring and facilitated navigation. (Adapted from Tidwell, J., *Designing Interfaces,* 2nd ed., O'Reilly Media, Sebastopol, California, 2010 [2].)

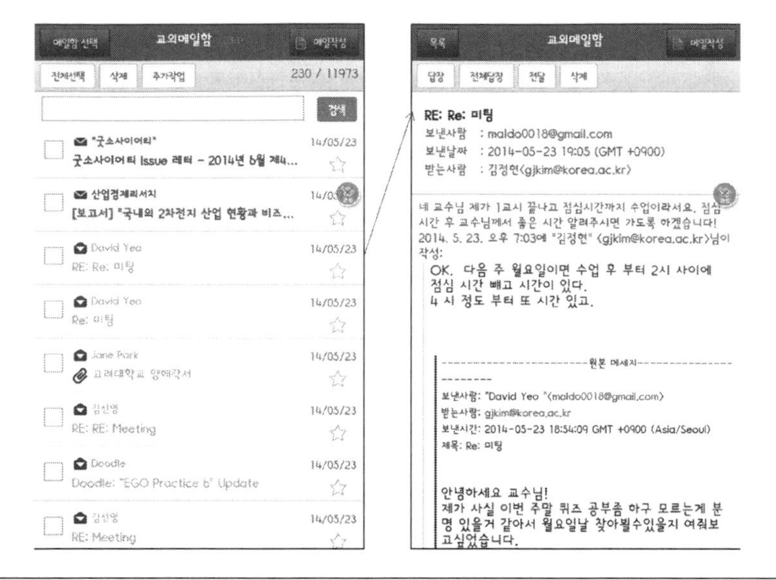

Figure 2.3 The use of one-window drilldown as a design pattern for content organization and fast navigation. (Adapted from Tidwell, J., *Designing Interfaces,* 2nd ed., O'Reilly Media, Sebastopol, California, 2010 [2].)

What:

> Show each of the application's pages within a single window. As a user drills down through a menu of options, or into an object's details, replace the window contents completely with the new page.

Use when:

> Your application consists of many pages or panels of content for the user to navigate through.... For a device with tight space restrictions,...you may have a complexity limit. Your users [also] may not be habitual computer users—having many application windows open at once may confuse them.

2.2.3 Taking User Input (General HCI Design)

Clever designs for taking user input (e.g., raw information or system commands) can improve the overall performance, in terms of both time and accuracy, for highly interactive systems. Modern interfaces employ graphical user interface (GUI) elements (e.g., window, text box, button, menu, forms, dialog box, icon), support techniques

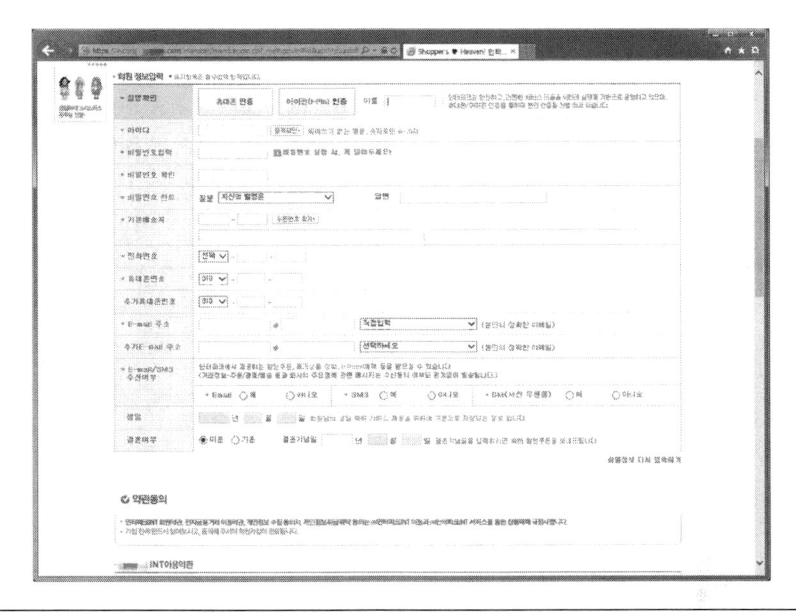

Figure 2.4 Display layout and user interfaces for facilitated date entry: Selection menus, default values, and structured forms are used to reduce errors. (From Smith, S. L., and Mosier, J. N., *Guidelines for Designing User Interface Software,* Mitre Corporation, Bedford, MA, 1986 [4].)

(e.g., autocompletion, deactivating irrelevant options, voice recognition), and devices (e.g., mouse, touch screen) to obtain user input in different ways. It is up to the UI designer to compose these input methods for the best performance with respect to the design constraints (e.g., user type, task characteristics, operating environment, etc.). Figure 2.4 is a collection of guidelines for use in applying these input methods to facilitate data entry [4].

1. *Consistency of data-entry transactions*: Similar sequences of actions should be used under all conditions (similar delimiters, abbreviations, etc.)
2. *Minimal input actions by user*: Fewer input actions means greater operator productivity. Make proper use of single-key commands, mouse selection, auto-completion features, and automatic cursor placement rather than typing/pressing in the full alphanumeric input. Selection from a list (e.g., by a menu or by mutually exclusive radio buttons) also reduces possibilities of error. Avoid switching between the keyboard and the mouse. Use default values.
3. *Minimal memory load on users*: When doing data entry, use menus and button choices so that users do not have to

remember a lengthy list of codes and complex syntactic command strings.

4. *Compatibility of data entry with data display*: The format of data-entry information should be linked closely to the format of displayed information (i.e., what you see is what you get).
5. *Clear and effective labeling of buttons and data-entry fields*: Use consistent labeling. Distinguish between required and optional data entry. Place labels close to the data-entry field.
6. *Match and place the sequence of data-entry and selection fields in a natural scanning and hand-movement direction* (e.g., top to bottom, left to right).
7. *Do not place semantically opposing entry/selection options close together*: For example, do not place "save" and "undo" buttons close together. Such a placement is likely to produce frequent erroneous input.
8. *Design of form and dialog boxes*: Most visual-display layout guidelines also apply to the design of form and dialog boxes.

Note that most of these guidelines apply only when using mouse/keyboard-driven GUI elements. Situations become more complicated when other forms of input are also used, such as touch, gesture, three-dimensional (3-D) selection, and voice. There are separate guidelines for incorporating such input modalities.

2.2.4 Users with Disability (User Type)

The W3C has led the Web Accessibility Initiative and published the *Web Content Accessibility Guidelines (WCAG) 2.0* [5]. It explains how to make web content more accessible to people with disabilities. Web content generally refers to the information in a web page or web application, including text, images, forms, sounds, and such (Figure 2.5). The following is a summary of the guidelines:

1. Perceivable
   - A. Provide text alternatives for nontext content.
   - B. Provide captions and other alternatives for multimedia.
   - C. Create content that can be presented in different ways, including by assistive technologies, without losing meaning.
   - D. Make it easier for users to see and hear content.

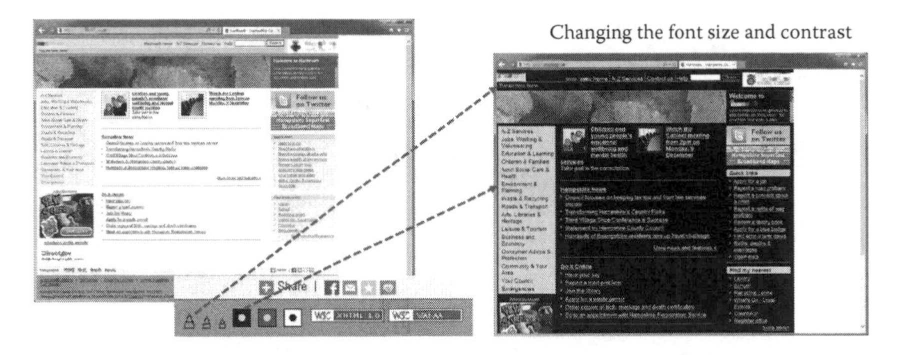

Figure 2.5 Adjustment feature for visually challenged users. The colors of the background and foreground text can be changed.

2. Operable
   A. Make all functionality available from a keyboard.
   B. Give users enough time to read and use content.
   C. Do not use content that causes seizures.
   D. Help users navigate and find content.
3. Understandable
   A. Make text readable and understandable.
   B. Make content appear and operate in predictable ways.
   C. Help users avoid and correct mistakes.
4. Robust
   A. Maximize compatibility with current and future user tools.

2.2.5 Mobile Device (Platform Type)

Recently, with the spread of smartphones, usability and user experience of mobile devices and applications has become even more important. Many conventional principles equally apply to mobile networked devices (Figure 2.6), but the following are more specific and important, as summarized by Tidwell [2]:

1. Fast status information (especially with regard to network connection and services)
2. Minimize typing and leverage on varied input hardware (e.g., buttons, touch, voice, handwriting recognition, virtual keyboard, etc.)
3. Fierce task focus (for less confusion in a highly dense information space)

Figure 2.6 Comparison of two mobile game interfaces (the initial entry screen): (a) information and object density is needlessly high and distracting (left), (b) simple and minimal layout, and object sizes fitted to ergonomic usage (right). (From http://www.withhive.com.)

4. Large hit targets (for easy and correct selection and manipulation)
5. Efficient use of screen space (with condensed information)

Following is a similar set of guidelines available from the Nokia developer's home page [6]:

1. Enable shortcuts (e.g., hot keys) for frequently used functions
2. Keep the user informed of his or her actions
3. Follow the device's (vendor's) interface patterns (positioning of the buttons and menus).

Figure 2.7 shows another design pattern put forth by Google® for the Android mobile interface [7]. It concerns the limited and different sizes of a family of handheld devices (i.e., smartphones, padlike devices, mobile Internet devices, netbooks) and more specifically

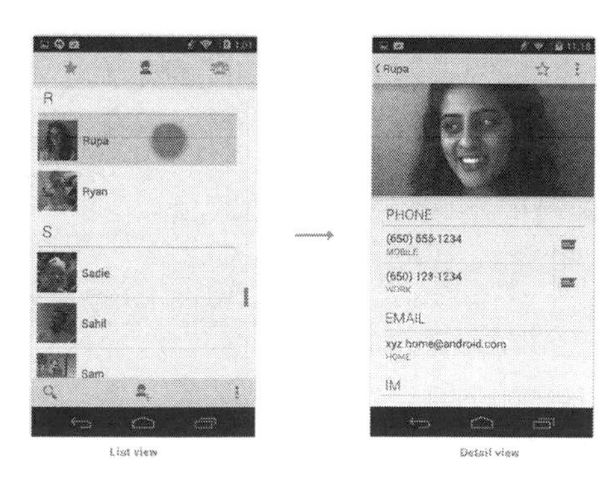

Figure 2.7 Android design guideline promoting the use of list views and detailed views (multiple panels) to efficiently use the screen size of mobile devices. (From Google, Multi-Pane Layouts, 2013, http://developer.android.com/design/patterns/multi-pane-layouts.html.)

suggests the use of "panels" as a way to achieve usability under such hardware constraints.

> Make sure that your app consistently provides a balanced and aesthetically pleasing layout by adjusting its content to varying screen sizes and orientations....
>
> ...Panels are a great way for your app to achieve this. They allow you to combine multiple views into one compound view when a lot of horizontal screen real estate is available and by splitting them up when less space is available.

2.2.6 Icons for Apple® iOS and Fonts for Windows® XP (Vendor)

Major vendors publish style guides for user-interaction elements to be used for applications running on their platform. For instance, Apple has published a design guideline document [8] that details how application icons should be designed and stylized:

1. Try to balance eye appeal and clarity of meaning in your icon so that it is rich and beautiful and clearly conveys the essence of your app's purpose.
2. Investigate how your choice of image and color might be interpreted by people from different cultures.
3. Create different sizes of your app icon for different devices. For iPhone and iPod touch, both of these sizes are required: (a) iPhone: 57 × 57 pixels and 114 × 114 pixels (high resolution) and (b) iPad: 72 × 72 pixels and 144 × 144 (high resolution). When iOS displays the app icon on the home screen of a device, it automatically adds the following visual effects: (a) rounded corners, (b) drop shadow, and (c) reflective shine.

Another example is the suggested choice of fonts/sizes for Windows XP or applications based on it [9]. These guidelines promote organizational styling and its identity and, ultimately, its consistency in user interfaces.

1. Franklin Gothic is used only for text over 14-point size. It is used for headers and should never be used for body text.
2. Tahoma is used as the system's default font. Tahoma should be used at 8-, 9-, or 11-point sizes.

Figure 2.8 An example of Trebuchet font used for a window title bar. (From Microsoft®, Windows XP Design Guidelines, 2002, http://msdn.microsoft.com/en-us/library/windows/hardware/gg463466.aspx [10].)

3. Verdana (bold, 8 point) is used only for title bars of tear-off/floating palettes.
4. Trebuchet MS (bold, 10 point) is used only for the title bars of Windows (Figure 2.8).

2.2.7 "Earcon" Design for Aural Interface (Modality)

Blattner, Sumikawa, and Greenberg [10] have suggested a few guidelines for designing "auditory" analog-to-visual icons. Similar to visual icons, which must capture the underlying meaning (for whatever it is trying to represent) and draw attention for easy recognition, earcons should be designed to be intuitive. They suggest three types of earcons, namely, those that are (a) symbolic, (b) nomic, and (c) metaphoric. Symbolic earcons rely on social convention such as applause for approval; nomic ones are physical such as a door slam; and metaphorical ones are based on capturing the similarities such as a falling pitch for a falling object [10]. Aural feedback (including earcons) involves a careful choice of sound-related parameters such as the amplitude/loudness, frequency/pitch, timbre, and duration. We take a more in-depth look at the aural modality in Chapter 3.

2.2.8 Cell Phones (or Making Calls) in Automobiles (Task)

Green et al. [11] have categorically outlined interface guidelines for automobiles and vehicles whose interfaces are nowadays mostly electronic and computer controlled, as seen in Table 2.3. The categories include design guidelines for manual control, spoken input and output, visual and auditory display, navigation guide, and cell phone consideration, to name just a few (Figure 2.9).

Table 2.3 Samples of Guidelines for Car Phone Interfaces in Vehicles

SUBCATEGORIES	GUIDELINE
Basic	Car phones should operate like phones people have at home. The use of *send* to make a connection and *power* to turn a phone on and off are notable inconsistencies.
Voice dialog	Verbal commands and button labels should use the same terms. Commands of interest include *dial*, *store*, *recall*, and *clear*. This is an instance of the consistency principle.
Manual dialing	The *store* and *recall* buttons, used for similar functions, should be adjacent to each other. This is an instance of the grouping principle.

Source: Green, P., Levison, W., Paelke, G., and Serafin, C. *Suggested Human Factors Design Guidelines for Driver Information Systems*, Technical Report UMTRI-93-21, Transportation Research Institute, University of Michigan, Ann Arbor, 1993 [11].

Figure 2.9 Phone interface for automobiles.

2.2.9 E-Commerce (Application)

Kalsbeek [12] has collected and formulated very extensive, detailed, and structured HCI guidelines for e-commerce applications. A total of 404 guidelines structured in four groups (general, input/output forms, UI elements, and checkout process) are given and applied to several real systems for validation and evaluation. The following is a guideline under the checkout-process section concerning the steps of a subtask (the checkout process).

> Check-out should start at the shopping cart, followed by the gift options or shipping method, the shipping address, the billing address, payment information, order review and finally an order summary.... Then the

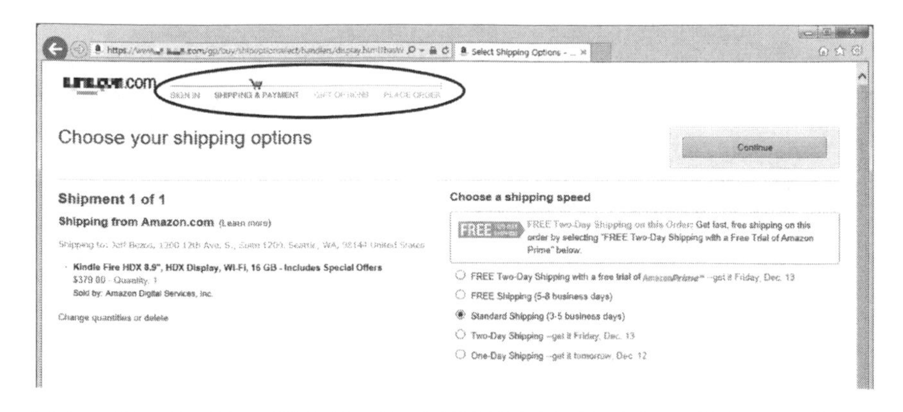

Figure 2.10 Status information (circled) shown in the process of a book purchase at Amazon.com.

site displays a confirmation page and gives customers the option to register. The checkout process is linear.

Figure 2.10 shows the status information (circled) shown in the process of a book purchase at Amazon.com.

2.3 Summary

While most of the guidelines—specific or general—seem quite straightforward and are easy to understand, incorporating them in actual design and implementation is very difficult. Many guidelines are still at quite a high level, similar to the HCI principles, and leave the developer wondering how to actually apply them in practice. Another reason is that there are just too many different aspects to consider (especially for a large-scale system). Sometimes, the guidelines can even be in conflict with each other, which requires prioritizing on the part of the designer. For instance, it can be difficult to give contrast to an item for highlighting its importance when one is restricted to using certain colors, e.g., for a corporate identity purpose. Another example might be when attempting to introduce a new interface technology (e.g., touch gestures). While the new interface may have been proven effective in the laboratory, it still may require significant familiarizing and training on the part of the user. It is often the case that external constraints such as monetary and human resources restrict sound HCI practice.

There is no straight answer to how such conflicts can be managed and how to incorporate all the requirements simultaneously, particularly under stringent external constraints. One must realize that all designs involve compromises and tradeoffs. Experienced designers understand the ultimate benefit and cost for practicing sound HCI design. In spite of the acknowledged aspect of "black art" to HCI design (in which good judgments are made by experienced developers), the HCI guidelines still help greatly to ensure overall usability and performance. In Chapter 3, we will study cognitive and ergonomic knowledge (more theoretical), which, along with the principles and guidelines we have learned so far (more experiential), will be applied to HCI design.

References

1. Wikipedia. 2013. ISO 9241. http://en.wikipedia.org/wiki/ISO_9241.
2. Tidwell, Jennifer. 2010. *Designing interfaces*. 2nd ed. Sebastopol, CA: O'Reilly Media.
3. Leavitt, Michael O., and Ben Shneiderman. 2006. *Research-based web design and usability guidelines*. Washington, DC: US Department of Health and Human Services.
4. Smith, Sidney L., and Jane N. Mosier. 1986. *Guidelines for designing user interface software*. Bedford, MA: Mitre Corporation.
5. Caldwell, Ben, Michael Cooper, Loretta G. Reid, and Gregg Vanderheiden, eds. 2010. *Web content accessibility guidelines (WCAG) 2.0*. W3C. http://www.w3.org/WAI/GL/WCAG20/.
6. Nokia. 2012. Guidelines for mobile interface design. http://www.developer.nokia.com/Community/Wiki/Guidelines_for_Mobile_Interface_Design.
7. Android. 2013. Multi-pane layouts. http://developer.android.com/design/patterns/multi-pane-layouts.html.
8. Apple. 2014. iOS human interface guidelines. http://developer.apple.com/library/ios/documentation/userexperience/conceptual/mobilehig/MobileHIG.pdf.
9. Microsoft (Windows XP Design Team). 2001. Windows XP visual guidelines. Microsoft Corporation.
10. Blattner, Meera M., Denise A. Sumikawa, and Robert M. Greenberg. 1989. Earcons and icons: Their structure and common design principles. *Human–Computer Interaction* 4 (1): 11–44.
11. Green, Paul, William Levison, Gretchen Paelke, and Colleen Serafin. 1993. *Suggested human factors design guidelines for driver information systems*. Technical Report UMTRI-93-21. Ann Arbor: University of Michigan, Transportation Research Institute.

12. Kalsbeek, Maarten. 2012. Interface and interaction design patterns for e-commerce checkouts. Master's thesis, University of Twente. http://essay.utwente.nl/62507/.

3
Human Factors as HCI Theories

3.1 Human Information Processing

Any effort to design an effective interface for human–computer interaction (HCI) requires two basic elements: an understanding of (a) computer factors (software/hardware) and (b) human behavior. We will look at the computer aspects of HCI design in the second part of this book. In this chapter, we take a brief look at some of the basic human factors that constrict the extent of this interaction.

In Chapters 1 and 2, we studied two bodies of knowledge for HCI design, namely (a) high-level and abstract principles and (b) specific HCI guidelines. To practice user-centered design by following these principles and guidelines, the interface requirements must often be investigated, solicited, derived, and understood directly from the target users through focus interviews and surveys. However, it is also possible to obtain a fairly good understanding of the target user from knowledge of human factors. As the main underlying theory for HCI, human factors can largely be divided into: (a) cognitive science, which explains the human's capability and model of conscious processing of high-level information and (b) ergonomics, which elucidates how raw external stimulation signals are accepted by our five senses, are processed up to the preattentive level, and are later acted upon in the outer world through the motor organs. Human-factors knowledge will particularly help us design HCI in the following ways.

- *Task/interaction modeling*: Formulate the steps for how humans might interact to solve and carry out a given task/problem and derive the interaction model. A careful HCI designer would not neglect to obtain this model by direct observation of the users themselves, but the designer's knowledge in cognitive science will help greatly in developing the model.

- *Prediction, assessment, and evaluation of interactive behavior*: Understand and predict how humans might react mentally to various information-presentation and input-solicitation methods as a basis for interface selection. Also, evaluate interaction models and interface implementations and explain or predict their performance and usability.

3.1.1 Task Modeling and Human Problem-Solving Model

The HCI principle of task/interaction modeling was helpful in understanding the tasks required to accomplish the ultimate goal of the interactive system. For instance, a goal of a word-processing system might be to produce a nice-looking document as easily as possible. In more abstract terms, this whole process of interaction could be viewed as a human attempting to solve a "problem" and applying certain "actions" on "objects" to arrive at a final "solution." Cognitive science has investigated the ways in which humans solve problems, and such a model can help HCI designers analyze the task and base the interaction model or interface structure around this innate problem-solving process. Thus for a smaller problem of "fixing the font," the action could be a "menu item selection" applied to a "highlighted text." There are several "human problem-solving" models that are put forth by a number of researchers, but most of them can be collectively summarized as depicted in Figure 3.1. This problem-solving process epitomizes the overall information-processing model. In general, human problem-solving or information-processing efforts consist of these important parts:

- *Sensation*, which senses external information (e.g., visual, aural, haptic), and *Perception*, which interprets and extracts basic meanings of the external information. (As a lower level part of the information-processing chain [more ergonomic], we take a closer look at these and how they relate to HCI in Section 3.2.)
- *Memory*, which stores momentary and short-term information or long-term knowledge. This knowledge includes

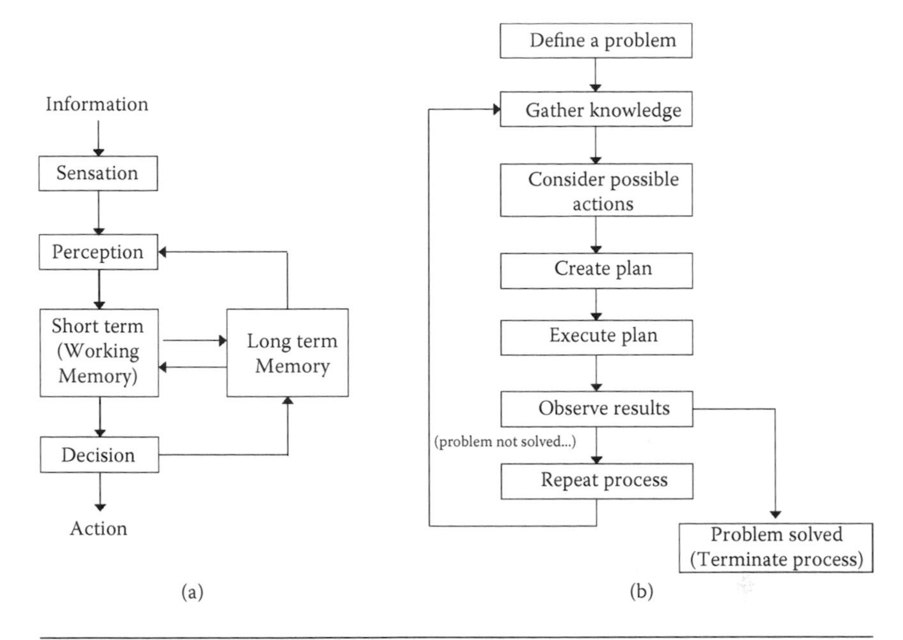

Figure 3.1 (a) The overall human problem-solving model and process and (b) a more detailed view of the "decision maker/executor."

information about the external world, procedures, rules, relations, schemas, candidates of actions to apply, the current objective (e.g., accomplishing the interactive task successfully), the plan of action, etc.

- *Decision maker/executor*, which formulates and revises a "plan," then decides what to do based on the various knowledge in the memory, and finally acts it out by commanding the motor system (e.g., to click the mouse left button).

Figure 3.1b shows the overall process in a flowchart. Once a problem is defined and identified as one that needs to be solved (simply by the user's intention), it is established as the top goal. Then a hierarchical plan (Figure 3.2) is formulated by refining the goal into a number of subgoals. A number of actions or subtasks are identified in the hope of solving the individual subgoals considering the external situation. By enacting the series of these subtasks to solve the subgoals, the top goal is eventually accomplished. Note that enacting the subtasks does not guarantee their successful completion (i.e., they may fail). Thus the whole process is repeated by observing the resulting situation and revising and restoring the plan.

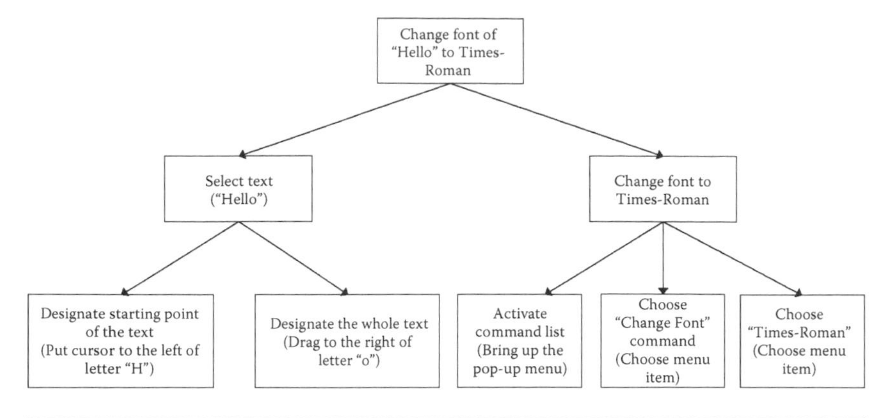

Figure 3.2 An example of a hierarchical task model of changing a font for a short text. Note that a specific interface may be chosen to accomplish the subtasks in the bottom.

Figure 3.2 shows an example of a hierarchical task plan (equivalent to hierarchical goal structure) illustrating how the simple task of changing the font of a text could be solved, i.e., what kinds of basic tasks would be needed. Note that in a general hierarchical task model, certain subtasks need to be applied in series, and some may need to be applied concurrently. One can readily appreciate from the simple example in Figure 3.2 how an interactive task model can be hierarchically refined and can serve as a basis for the interface structure. Note that, based on this model, we could "select" interfaces to realize each subtask in the bottom of the hierarchy, which illustrates the crux of the HCI design process. The interaction model must represent as much as possible what the user has in mind, especially what the user expects must be done (the mental model) in order to accomplish the overall task. This way, the user will be "in tune" with the resulting interactive application. The interface selection should be done based on ergonomics, user preference, and other requirements or constraints. Finally, the subtask structure can lend itself to the menu structure, and the actions and objects to which the actions apply can serve as the basis for an object-class diagram (for an object-oriented interactive software implementation).

3.1.2 Human Reaction and Prediction of Cognitive Performance

We can also, to some degree, predict how humans will react and perform in response to a particular human interface design. We can consider two aspects of human performance: one that is cognitive and the other ergonomic. In the remainder of this section and in Section 3.2, we focus on the cognitive aspects. Ergonomic aspects are discussed in Section 3.3.

Norman and Draper [1] spoke of the "gulf of execution/evaluation," which explains how users can be left bewildered (and not perform very well) when an interactive system does not offer certain actions or does not result in a state as expected by the user (Figure 3.3). Such a phenomenon would be a result of an interface based on an ill-modeled interaction. A user, when solving a problem or using an interactive system to do so, will first form a mental model that is mostly equivalent to the hierarchical "action" plan for the task (see Section 3.1.1). The mismatch between the user's mental model and the task model employed by the interactive system creates the "gulf." On the other hand, when the task model and interface structure of the interactive system maps well to the expected mental model of the user, the task performance will be very fluid.

Memory capacity also influences the interactive performance greatly. As shown in Figure 3.1, there are largely two types of memory in the human cognitive system: the short term and the long term. The short-term memory is also sometimes known as the working memory, in the sense that it contains (changing) memory elements meaningful for the task at hand (or chunks). Humans are known to remember about eight chunks of memory lasting only a very short amount of time [2]. This means that an interface cannot rely on the human's

Figure 3.3 Gulf of execution and evaluation: the gap between the expected and actual.

Figure 3.4 A snapshot of an online shopping process that does not display superfluous user status that can lead to anxiety, uncertainty, and erroneous response.

short-term memory beyond this capacity for fast operation. Imagine an interface with a large number of options or menu items. The user would have to rescan the available options a number of times to make the final selection. In an online purchasing system, the user might not be able to remember all of the relevant information such as items purchased, delivery options, credit card chosen, billing address, usage of discount cards, etc. (Figure 3.4). Thus such information will have to be presented to the user from time to time to refresh one's memory and ensure that no errors are made.

Retrieving information from the long-term memory is a difficult and relatively time-consuming task. Therefore, if an interactive system (e.g., targeted even for experts) requires expert-level knowledge, it needs to be displayed so as to at least elicit "recognition" (among a number of options) of it rather than completely relying on recall from scratch.

Memory-related performance issues are also important in multitasking. Many modern computing settings offer multitasking environments. It is known that when the user switches from one task to another, a "context switch" occurs in the brain, which means that the working memory content is replaced (and stored back into the long-term memory) with chunks relevant for the switched task (such as the state of the task up to that moment). This process can bring about overall degradation in task performance in many respects [3]. For an individual application to help itself in its use during multitasking, it can assist the user's context-switch process by capturing the context

Figure 3.5 Reminding the user of the context for multitasking for fast application switching (top part of the figure).

information during its suspension, and by later displaying, reminding, and highlighting the information upon resumption (Figure 3.5).

3.1.2.1 Predictive Performance Assessment: GOMS Many important cognitive activities have been analyzed in terms of their typical approximate process time, e.g., for single-chunk retrieval from the short-term memory, encoding (memorizing) of information into the long-term memory, responding to a visual stimulus and interpreting its content, etc. [4–6]. Based on these figures and a task-sequence model, one might be able to quantitatively estimate the time taken to complete a given task and, therefore, make an evaluation with regard to the original performance requirements. Tables 3.1 and 3.2 illustrate such an example based on the framework called GOMS (Goals, Operators, Methods, and Selection) [7].

Table 3.1 Estimates of Time Taken for Typical Desktop Computer Operations from GOMS

TYPE OF OPERATION	TIME ESTIMATE
K: Keyboard input	Expert: 0.12 s Average: 0.20 s Novice: 1.2 s
T(*n*): Type *n* characters	280 × *n* ms
P: Point with mouse to something on the display	1100 ms
B: Press or release mouse button	100 ms
BB: Click a mouse button (press and release)	200 ms
H: Home hands, either to the keyboard or mouse	400 ms
M: Thinking what to do (mental operator)	1200 ms (can change)
W(*t*): Waiting for the system (to respond)	*t* ms

Source: Card, S. K., Moran, T. P., and Newell, A., The Model Human Processor: An Engineering Model of Human Performance, in *Handbook of Human Perception*, vol. 2, *Cognitive Processes and Performance*, ed. K. R. Boff, L. Kauffman, and J. P. Thomas, 1–35, John Wiley and Sons, New York, 1986 [7].

Table 3.2 Estimates of Time Taken for Two Task Models of "Deleting a File"

DELETING A FILE			
DESIGN 1		DESIGN 2[a]	
1. Point to file icon	P	1. Point to file icon	P
2. Click mouse button	BB	2. Click mouse button	BB
3. Point to file menu	P	3. Move hand to keyboard	M
4. Press and hold mouse button	B	4. Hit command key: command-T	KK
5. Point to DELETE item	P	5. Move hand back to mouse	H
6. Release mouse button	B		
7. Point to original window	P		
Total time = 4.8 s		Total time = 2.66 s	

Note: The total time is computed by adding the corresponding figures in Table 3.1.
[a] Design 2 is the "expert" version that uses a hot key [7].

The GOMS evaluation methodology starts by the same hierarchical task modeling we have described in Section 3.1. Once a sequence of subtasks is derived, one might map a specific operator in Table 3.1 (or, in other words, interface) to each of the subtasks. With the preestablished performance measures (Table 3.1), the total time of task performance can be easily calculated by summing the task times of the whole set of subtasks. Different operator mappings can be tried comparatively in terms of their performance. The original GOMS model was developed mainly for the desktop computing environment, with performance figures for mouse clicks, keyboard input, hand movement, and mental operators (Table 3.1). Even though this model was created nearly 30 years ago, the figures are still amazingly valid. (While computer technologies have advanced much since then, humans' capabilities have remained mostly the same.) GOMS models for other computing environments have been proposed as well [8].

Table 3.1 shows different performance measures for various task operators or interfaces [7]. Table 3.2 shows two designs of the main task of "file deletion." Each design is decomposed in a slightly different manner and with operators mapped to the individual subtasks, resulting in different total times of operation (the first in 4.8 s and the second in 2.7 s). GOMS is quite simple in that it can only evaluate in terms of the task performance, while there are many other criteria by which an HCI design should be evaluated. Moreover, among the operators, the mental operator approximates the time taken for "momentary thought or memory retrieval" in between motor tasks (like mouse clicks). Obviously, there can be some inaccuracies introduced in the use of the mental operators during the interaction modeling process.*

3.2 Sensation and Perception of Information

The previous section explained the value and usage of the knowledge of cognitive and high-level information processing to HCI design. We now shift our focus to raw information processing. First we look at the input side (i.e., the human sensory system). Humans are known to have at least five senses. Among them, those that would be relevant to HCI (at least for now) are the modalities of visual, aural, haptic (force feedback), and tactile sensation. Taking external stimulation or raw sensory information (sometimes computer generated) and then processing it for perception is the first part in any human–computer interaction. Naturally, the information must be supplied in a fashion that is amenable to human consumption, that is, within the bounds of a human's perceptual capabilities.

Another aspect of sensation and perception is *attention*, that is, how to make the user selectively (consciously or otherwise) tune in to a particular part of the information or stimulation. Highly attentive

* The developers of GOMS do outline a strategy for when to properly use the mental operators for a correct task modeling and performance prediction [7].

information can be used for alerts, reminders, highlighting of prioritized/structured information, guidance, etc. Note that attention must occur and be modulated within *awareness* of the larger task(s). While we might tune in to certain important information, we often still need to have an understanding, albeit approximate, of the other activities or concurrent tasks, such as in multitasking or parallel processing of information.

In the following discussion, we examine the processes of sensation and perception in the four major modalities and the associated human capabilities in this regard. Just as cognitive science was useful in interaction and task modeling, this knowledge is essential in sound interface selection and design.

3.2.1 Visual

Visual modality is by far the most important information medium. Over 40% of the human brain is said to be involved with the processing of visual information. As already mentioned, the parameters of the visual interface design and display system will have to conform to the capacity and characteristics of the human visual system. In this section, we review some of the important properties of the human visual system and their implications for interface design. First we take a look at a typical visual interaction situation as shown in Figure 3.6.

3.2.1.1 Visual and Display Parameters

- *Field of view (FOV)*: This is the angle subtended by the visible area by the human user in the horizontal or vertical direction. The shaded area in Figure 3.6 illustrates the horizontal field of view. The human FOV is nearly 180° in both the horizontal and vertical directions.
- *Viewing distance*: This the perpendicular distance to the surface of the display. Viewing distance (dotted line in Figure 3.6) may change with user movements. However, one might be able to define a nominal and typical viewing distance for a given task or operating environment.
- *Display field of view*: This is the angle subtended by the display area from a particular viewing distance. Note that for the same fixed display area, the display FOV will be different at

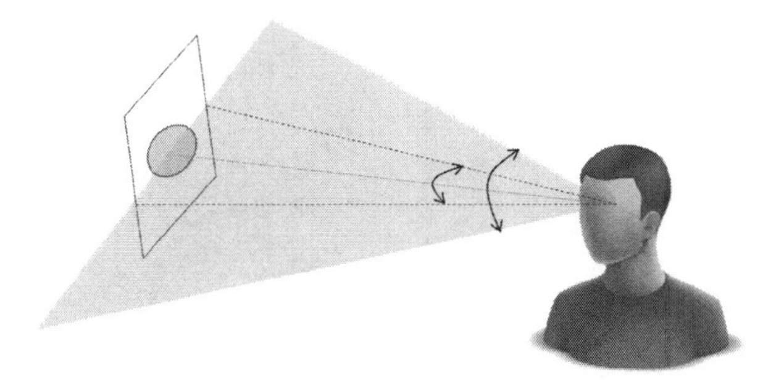

Figure 3.6 A user viewing a display system. The shaded area illustrates the horizontal field of view (shown to be much less than the actual for illustration purpose), while the dashed line is the same as offered by the display. The display offers different fields of view depending on the viewing distance (dotted line in the middle). The oval shape in the display represents the approximate area for which high details are perceived through the corresponding foveal area in the user eyes.

different viewing distances. In Figure 3.6, the display FOV is denoted with the dashed line. The display offers different fields of view, depending on the viewing distance (dotted line in the middle).

- *Pixel*: A display system is typically composed of an array of small rectangular areas called *pixels*.
- *Display resolution*: This is the number of pixels in the horizontal and vertical directions for a fixed area.
- *Visual acuity*: In effect, this is the resolution perceivable by the human eye from a fixed distance. This is also synonymous with the power of sight, which is different for different people and age groups.

These human visual and display parameters need to be matched as much as possible to provide a comfortable and effective visual display environment, for instance, display FOV to human FOV, display resolution/object size to visual acuity, and so forth (Figure 3.7). Note that the display FOV is more important than the absolute size of the display. A distant large display can have the same display FOV as a close small display, even though it may incur different viewing experiences. If possible, it is desirable to choose the most economical display, not necessarily the biggest or the one with the highest resolution, with respect to the requirement of the task and the typical user characteristics.

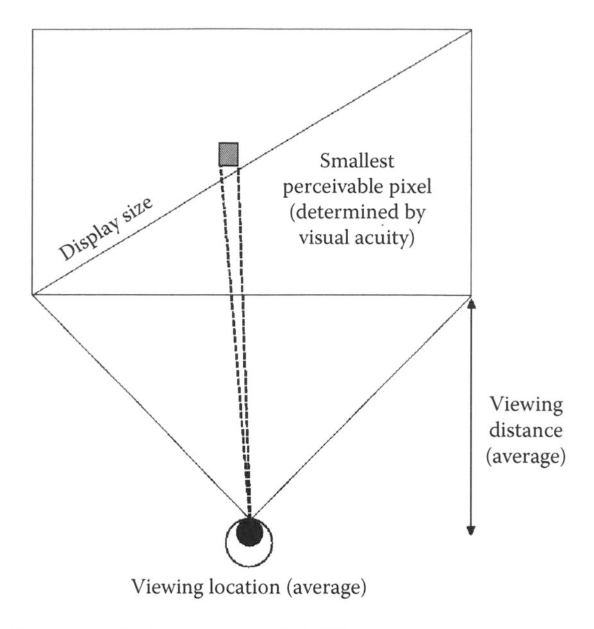

Figure 3.7 The display system parameters: display size, resolution, pixel determined by the user's visual acuity, and viewing location.

3.2.1.2 Detail and Peripheral Vision The human eye contains two types of cells that react to light intensities in different ways. The *cones*, which are responsible for color and detail recognition, are distributed heavily in the center of the retina (back of the eyeball), which subtends about 5° in the human FOV and roughly establishes the area of focus. The oval region in Figure 3.6 shows the corresponding region in the display for which details can be perceived through these cells. On the other hand, the *rods* are distributed mainly in the periphery of the retina and are responsible for motion detection and less detailed peripheral vision. While details may not be sensed, the rods contribute to our awareness of the surrounding environment.

Differently from that of human perception, most displays have uniform resolution. However, if the object details can be adjusted depending on where the user is looking or based on what the user may be interested in (Figure 3.8a), the overall rendering of the image can be made more economical or to doubly emphasize certain objects relative to others in their neighborhood (e.g., detail contrast). We may assess the utility of a large, very-high-resolution display system such as the one shown in Figure 3.9. From a nominal viewing distance, only

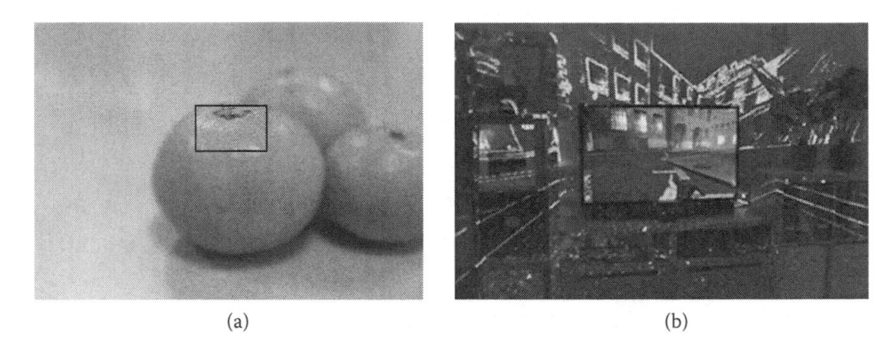

(a) (b)

Figure 3.8 (a) An ideal display that would provide relatively higher resolution in the area of the user's focus; (b) a large immersive display as realized by a high-resolution monitor in the middle- and lower-resolution projection in the periphery. (From Microsoft® Research, CHI 2013: An Immersive Event (Illusions create an immersive experience), 2013, http://research.microsoft.com/en-us/news/features/chi2013-042913.aspx [9].)

Figure 3.9 A large, tiled, high-resolution display. Is it really worth the cost? (From Ni, T., Schmidt, G. S., Staadt, O. G., Livingston, M. A., Ball, R., and May, R. A., *Proceedings of IEEE Virtual Reality Conference*, IEEE, Piscataway, NJ, 2006, pp. 223–236 [10].)

a small portion of the large display will correspond to the foveal area. Thus, while viewing one portion, the other major parts of the large-display resolution could go to "waste" (unless used by multiple users at once, as in an IMAX theater). Consequently, it can be argued that it is more economical to use a smaller high-resolution display placed at a close distance. Interestingly, Microsoft Research recently introduced a display system called the *Illumiroom* [9] in which a high-resolution display is used in the middle, and a wide low-resolution projection and peripheral display provides high immersion (Figure 3.8b).

3.2.1.3 Color, Brightness, and Contrast Other important properties and attributes of visual quality are brightness, color, and contrast.

- *Brightness*: The amount of light energy emitted by the object (or as perceived by the human).
- *Color*: Human response to different wavelengths of light, namely for those corresponding to red, green, blue, and their mixtures. A color can be specified by the composure of the amounts contributed by the three fundamental colors and also by hue (particular wavelength), saturation (relative difference in the major wavelength and the rest in the light), and brightness value (total amount of the light energy) (Figure 3.10).
- *Contrast*: Relative difference in brightness or color between two visual objects. Contrast in brightness is measured in terms of the difference or ratio of the amounts of light energies between two or more objects. The recommended ratio of the foreground to background brightness contrast is at least 3:1. Color contrast is defined in terms of differences or ratios in the dimensions of hue and saturation. It is said that the brightness contrast is more effective for detail perception than the color contrast (Figure 3.11).

3.2.1.4 Pre-Attentive Features and High-Level Diagrammatic Semantics Detail, color, brightness, and contrast are all very-low-level raw visual properties. Before all these low-level-part features are finally

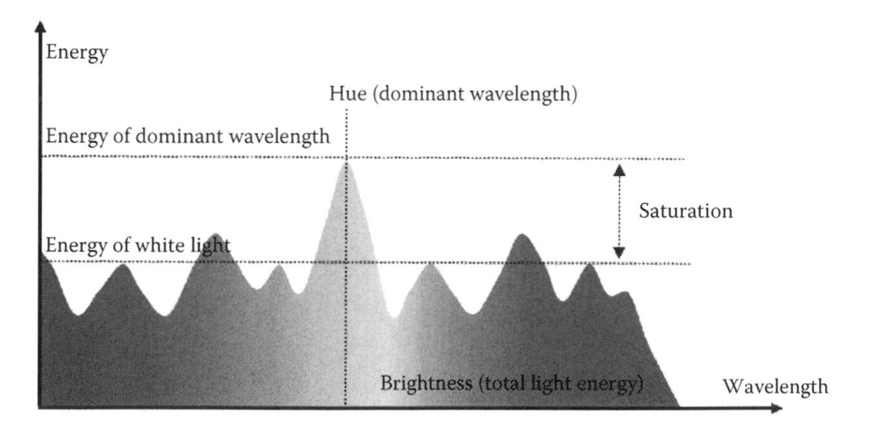

Figure 3.10 Color specification by hue (particular/dominant wavelength), saturation (relative difference in the major wavelength and the rest), and value/brightness (total amount of the light energy).

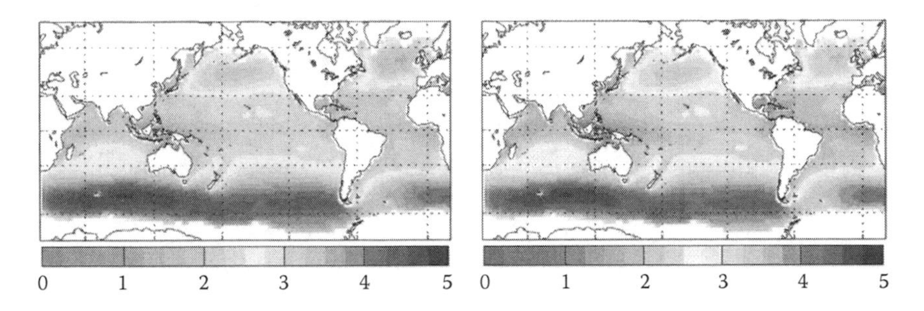

Figure 3.11 Coding of information in a map (e.g., temperature levels) using contrast in brightness (left) and color (right). (From Hemer, M. A., Fan, Y., Mori, N., Semedo, A., and Wang, X. L., *Nature Climate Change*, 3, 471–476, 2013 [11].)

consolidated for conscious recognition (of a larger object) through the visual information processing pipeline, *pre-attentive* features might be used to attract our attention. Pre-attentive features are composite, primitive, and intermediate visual elements that are automatically recognized before entering our consciousness, typically within 10 ms after entering the sensory system [12]. These features may rely on the relative differences in color, size, shape, orientation, depth, texture, motion, etc. Figure 3.12 shows several examples and how they can be used collectively to form and design effective graphic icons.

At a more conscious level, humans may universally recognize certain high-level complex geometric shapes and properties as a whole and understand the underlying concepts. Figure 3.13 shows examples of such universally accepted (across different cultures) geometric diagrams with

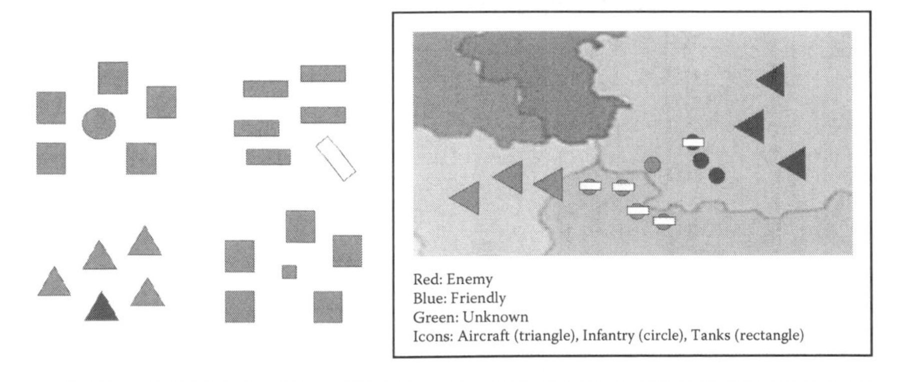

Figure 3.12 Examples of preattentive features for attention focus based on differences in size, shape, and orientation (left) and application to icon design (right). (From Ware, C., *Information Visualization: Perception for Design,* 3rd ed., Morgan Kaufmann, Waltham, MA, 2012 [12].)

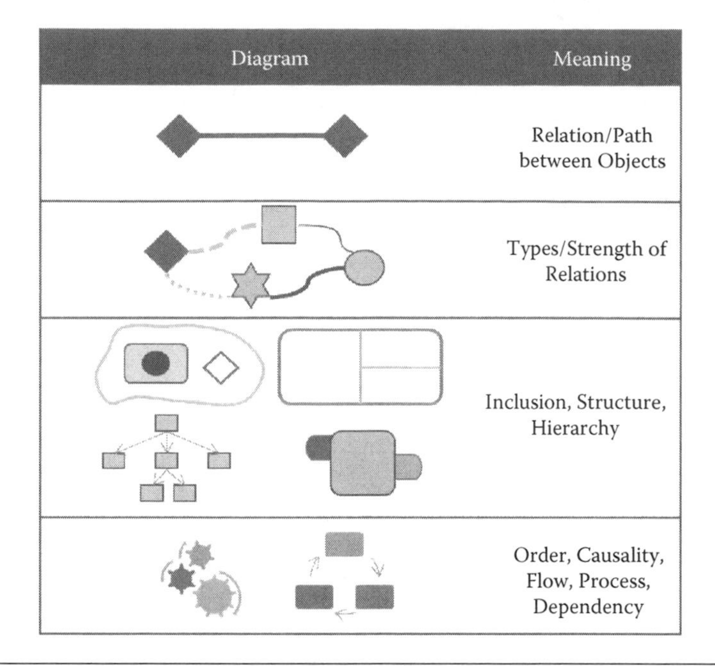

Figure 3.13 Examples of diagrams/shapes/objects/figures with universal semantics. (From Ware, C., *Information Visualization: Perception for Design*, 3rd ed., Morgan Kaufmann, Waltham, MA, 2012 [12].)

the connotation of, e.g., connection/relation, dependency, causality, inclusion, hierarchy/structure, flow/process, etc.

3.2.2 Aural

Next to the visual, the aural modality (sound) is perhaps the most prevalent mode for information feedback. The actual form of sound feedback can be roughly divided into three types: (a) simple beep-like sounds, (b) short symbolic sound bytes known as *earcons* (e.g., the paper-crunching sound when a file is inserted into the trashcan for deletion), and (c) relatively longer "as is" sound feedback that is replayed from recordings or synthesis. As we did for the visual modality, we will first go over some important parameters of the human aural capacity and the corresponding aural display parameters.

3.2.2.1 Aural Display Parameters

- *Intensity* (amplitude) refers to the amount of sound energy and is synonymous with the more familiar term, *volume*. Intensity is often measured in the units of decibels (dB), a logarithmic scale of sound energy, where 0 dB corresponds to the lowest level of audible sound and about 130 dB is the highest. It is instructive to know the decibel levels of different sounds as a guideline in setting the nominal volume for the sound feedback (Table 3.3).

- Sound can be viewed as containing or being composed of a number of sinusoidal waves with different *frequencies* and corresponding *amplitudes*. The dominant frequency components determine various characteristics of sounds such as the pitch (e.g., low or high key), timbre (e.g., which instrument), and even directionality (where is the sound coming from?). Humans can hear sound waves with frequency values between about 20 and 20,000 Hz [13].
- *Phase* refers to the time differences among sound waves that emanate from the same source. Phase differences occur, for example, because our left and right ears may have slightly different distances to the sound source and, as such, phase differences are also known to contribute to the perception of spatialized sound such as stereo.

When using aural feedback, it is important for the designer to set these fundamental parameters properly. A general recommendation is that the sound signal should be between 50 and 5000 Hz and composed of at least four prominent harmonic frequency components (frequencies that are integer multiples of one another), each within the range of 1000–4000 Hz [14]. Aural feedback is more commonly used in intermittent alarms. However, overly loud (i.e., needlessly high

Table 3.3 Examples of Different Sounds and Their Typical Intensity Levels in Decibels

INTENSITY (DB)	DESCRIPTION
0	Weakest sound audible
30	Whisper
50	Office environment
60	Normal conversation
110	Rock band
130	Pain threshold

amplitude) alarms are known to rather startle the user and lower the usability. Instead, other techniques can be used to attract attention and convey urgency by such aural feedback techniques as repetition, variations in frequency and volume, and gradual and aural contrast to the background ambient sound (e.g., in amplitude and frequency).

3.2.2.2 Other Characteristics of Sound as Interaction Feedback We further point out a few differences of aural feedback from the visual. First, sound is effectively omnidirectional. For this reason, sound is most often used to attract and direct a user's attention. However, as already mentioned, it can also be a nuisance as a task interrupter (e.g., a momentary loss of context) by the startle effect. Making use of contrast is possible with sound as well. For instance, auditory feedback would require a 15–30-dB difference from the ambient noise to be heard effectively. Differentiated frequency components can be used to convey certain information.

Continuous sound is somewhat more subject to becoming habituated (e.g., elevator background music) than stimulation with other modalities. In general, only one aural aspect can be interpreted at a time. That is, it is difficult to make out the aural content when the sound is jumbled/masked with multiple sources. Humans do possess an ability to tune in to a particular part of the sound (e.g., string section in a symphony); however, this requires much concentration and effort.

3.2.2.3 Aural Modality as Input Method So far, the aural modality has been explained only in the context of passive feedback. As for using it actively as a means for input to interactive systems, two major methods are: (a) keyword recognition and (b) natural language understanding.

Isolated-word-recognition technology (for enacting simple commands) has become very robust lately. In most cases, it still requires speaker-specific training or a relatively quiet background. Another related difficulty with voice input is the "segmentation" problem, i.e., how to segment out, from a stream of continuous voice input or background noise, the portion that corresponds to the actual command. As such, many voice input systems operate in an explicit mode or state. For example, the user has to press a button to activate the voice recognition (and enter into the recognition mode/state) and then speak the command into the microphone. (This also relieves the computational burden of having to run the voice-recognition process in the background if the system did not know when the command was to be heard.) The need to switch to the voice-command mode is still quite a nuisance to the ordinary user. Thus, voice input is much more effective in situations where, for example, hands are totally occupied or where modes are not necessary because there is very little background noise or because there is no mixture of conversation with the voice commands.

Machine understanding of long sentences and natural-language-based commands is still very computationally difficult and demanding. While not quite practical for everyday user-interface input methods, language-understanding technology is advancing fast, as demonstrated recently by the Apple® Siri [15] and IBM® Watson [16], where high-quality natural-language-understanding services are offered by the cloud (Figure 3.14). Captured segments of voice/text-input sentences can be sent to these cloud servers for very fast and near-real-time response. With the spread of smart-media client devices that might be computationally light yet equipped with a sleuth of sensors, such a cloud-based natural-language interaction (combined with intelligence) will revolutionize the way we interact with computers in the near future.

3.2.3 *Tactile and Haptic*

Interfaces with tactile and haptic feedback, while not yet very widespread, are starting to appear in limited forms. To be precise, the term

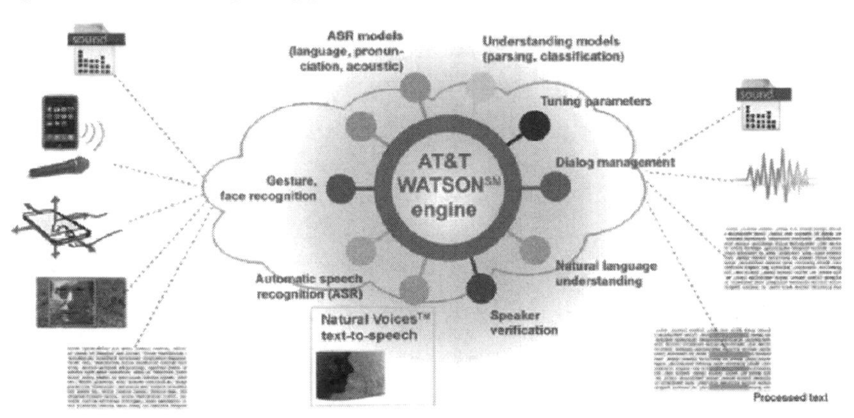

Figure 3.14 A high-quality natural-language-based interaction through the cloud. The smart-media client devices would send the captured sentence (in voice or text), and a correct and intelligent response is given back in real time. (From http://www.research.att.com/project/WATSON.)

haptic is defined to be the modality that takes advantage of touch by applying forces, vibrations, or motions to the user [17]. Thus *haptic* refers to both the sensation of force feedback as well as touch (tactile). For convenience, we will use the term *haptic* to refer to the modality for sensing force and kinesthetic feedback through our joints and muscles (even though any force feedback practically requires contact through the skin) and the term *tactile* for sensing different types of touch (e.g., texture, light pressure/contact, pain, vibration, and even temperature) through our skin.

3.2.3.1 Tactile Display Parameters

- *Tactile resolution*: The skin sensitivity to physical objects is different over the human body. The fingertip is one of the most sensitive areas and is frequently used for HCI purpose. The fingertip can sense objects as small as 40 μm in size [18].
- *Vibration frequency*: Rapid movement such as vibration is mostly sensed by the Pacinian corpuscle, which is known to have a signal-response range of 100–300 Hz. Vibration frequency of about 250 Hz is said to be the optimal for comfortable perception [16].
- *Pressure threshold*: The lightest amount of pressure humans can sense is said to be about 1000 N/m^2. For a fingertip, this amounts to about 0.02 N for the fingertip area [19]. The maximum threshold is difficult to measure, because when the force/torque gets large enough, the kinesthetic senses start to operate, and this threshold will greatly depend on the physical condition of the user (e.g., strong vs. weak user).

As mentioned previously, there are many types of tactile stimulation, such as texture, pressure, vibration, and even temperature. For the purposes of HCI, the following parameters are deemed important, and the same goes for the display system providing the tactile-based feedback. Physical tactile sensation is felt by a combination of skin cells and nerves tuned for particular types of stimulation, e.g., the Meissner's corpuscle for slight pressure or slow pushing (stimulation signal frequency of 3–40 Hz), Merkel cells for flutter and textured/protrusion surfaces (0.3–3 Hz), the Pacinian corpuscle for more rapid

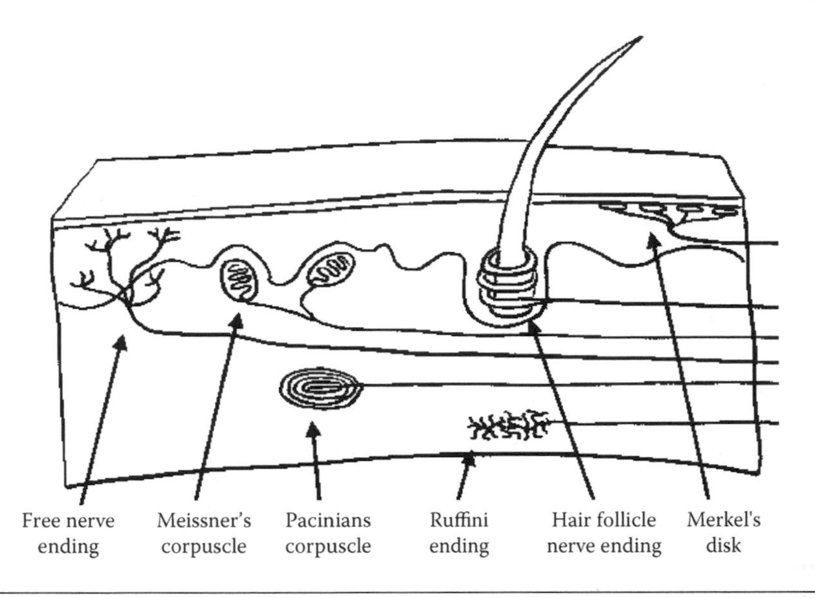

Figure 3.15 Cells and nerves in the skin. (From Proprioception, Intl. Encyclopedia of Rehabilitation, http://cirrie.buffalo.edu/encyclopedia/en/article/337.)

vibratory stimulation (10–500 Hz), and Ruffini endings for skin stretch (Figure 3.15).

While there are many research prototypes and commercial tactile display devices, the most practical one is the vibration motor, mostly applied in a single actuator configuration. Most vibration motors do not offer separate controllability for amplitude and frequency. In addition, most vibrators are not in direct contact with the stimulation target (e.g., the hand), making the signal somewhat muffled through the casing. Thus additional care is needed to set the right parameter values for the best effects under the circumstances.

Another way to realize vibratory tactile display is to use thin and light piezoelectric materials that exhibit vibration responses according to the amounts of electric potential supplied. Due to their flat form factor, such materials can be embedded, for instance, into flat touch screens. Sometimes sound speakers can be used to generate indirect vibratory feedback with high controllability (responding to wide ranges of amplitude and frequency signals) (Figure 3.16).

3.2.3.2 Haptic and Haptic Display Parameters Along with tactile feedback, haptic feedback adds a more apparent physical dimension to interaction. Force feedback and movement is felt by the cells and

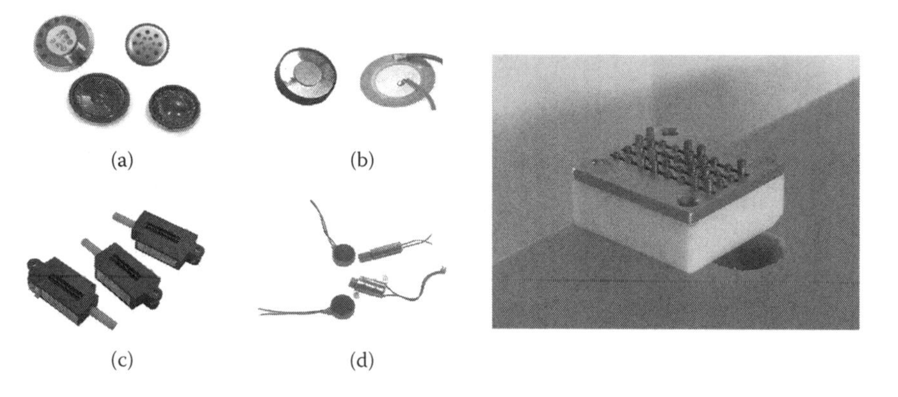

Figure 3.16 Left: Various actuators used for tactile feedback: (a) miniature speaker, (b) miniature electromagnet/latch, (c) piezoelectric strip, (d) microvibratory motors. Right: tactile array with multiple actuators. (From KU Leuven, Tactile Feedback, 2010, https://www.mech.kuleuven.be/en/pma/research/ras/researchtopics/tactfb.html [21].)

nerves in our muscles and joints. For instance, the muscle spindle/tendon takes the inertial load, and Pacinian/Ruffini/Golgi receptors sense the joint movements, pressure, and torque. The activation force for the joints is between 0.5 to 2.5 mN [20]. However, we can understand that this range would vary according to the user's age, gender, strength, size, weight, and so forth. Note that haptic devices are both input and output devices at the same time. (We briefly discuss this issue of haptic input in the next section in the context of human body ergonomics.)

The simplest form of a haptic device is a simple electromagnetic latch that is often used in game controllers. It generates a sudden inertial movement and slowly repositions itself for repeated usage. Normally, the user holds on to the device, and inertial forces are delivered in the direction relative to the game controller. Such a device is not appropriate for fast-occurring interaction (e.g., successive gun shots) or for displaying a continuously sustained force (e.g., leaning against a wall).

More-complicated haptic devices are in the form of a robotic kinematic chain, either fixed on the ground or worn on the body. As a kinematic chain, such devices offer higher degrees of freedom and finer force control (Figure 3.17). For the grounded device, the user interacts with the tip of the robotic chain through which a force feedback is delivered. The sensors in the joints of the device make it possible to track the tip (interaction point) within the three-dimensional (3-D) operating space.

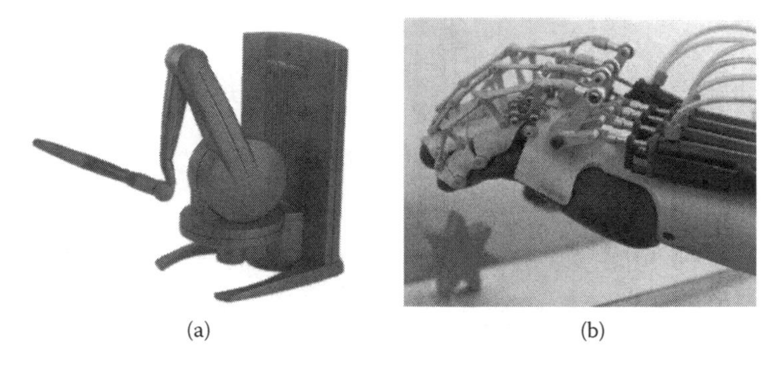

(a) (b)

Figure 3.17 Two types of haptic systems: (a) grounded and (b) body worn.

Using a similar control structure, body-worn devices transfer force with its mechanism directly attached to the body.

Important haptic display parameters are (a) the *degrees of freedom* (the number of directions in which force or torque be can displayed), (b) the *force range* (should be at least greater than 0.5mN), (c) *operating/interaction range* (how much movement is allowed through the device), and (d) *stability* (how stable the supplied force is felt to be). Stability is in fact a by-product of the proper *sampling period*, which refers to the time taken to sense the current amount of force at the interaction point and then determine whether the target value has been reached and reinforce it (a process that repeats until a target equilibrium force is reached at the interaction point). The ideal sampling period is about 1000 Hz, and when the sampling period falls under a certain value, the robotic mechanism exhibits instability (e.g., exhibited in the form of vibration) and thus lower usability. The dilemma is that providing a high sampling rate requires a heavy computation load, not only in updating the output force, but also in physical simulation (e.g., to check if the 3-D cursor has hit any virtual object). Therefore, a careful "satisficing" solution is needed to balance the level of the haptic device performance and the user experience (Figure 3.18).

In general, due to their mechanical nature, "robotic" haptic devices are not yet very practical. They tend to be heavy, clunky, dangerous, and take up a large volume. The cost is very high, often with only a small operating range, force range, or limited degrees of freedom. In many cases, simpler devices, such as one-directional latches or vibrators, are used in combination with visual and aural feedback to enrich the user experience.

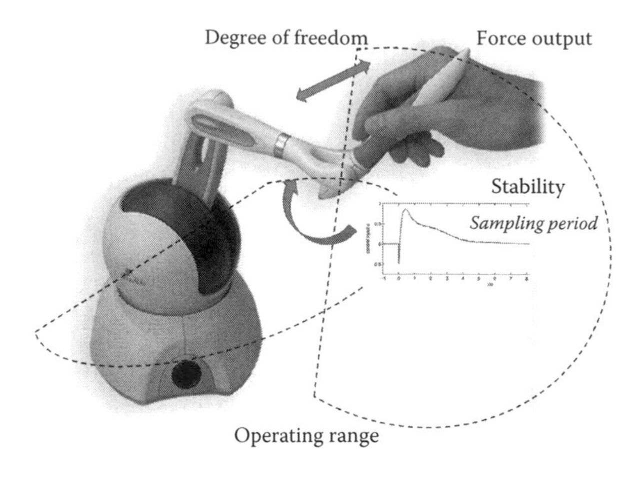

Figure 3.18 Important parameters for a haptic display system.

3.2.4 Multimodal Interaction

Conventional interfaces have been mostly visually oriented. However, for various reasons, multimodal interfaces are gaining popularity with the ubiquity of multimedia devices. By employing more than one modality, interfaces can become more effective in a number of ways, depending on how they are configured [22]. Here are some representative examples.

- *Complementary*: Different modalities can assume different roles and act in a complementary fashion to achieve specific interaction objectives. For example, an aural feedback can signify the arrival of a phone call while the visual displays the caller's name.
- *Redundant*: Different modality input methods or feedback can be used to ensure a reliable achievement of the interaction objective. For instance, the ring of a phone call can be simultaneously aural and tactile to strengthen the pick-up probability.
- *Alternative*: Providing users with alternative ways to interact gives people more choices. For instance, a phone call can be made either by touching a button or by speaking the callee's name, thereby promoting convenience and usability.

For multimodal interfaces to be effective, each feedback must be properly synchronized and consistent in its representation. For instance, to signify a button touch, the visual highlighting and beep sound effect must occur within a short time (e.g., less than 200 ms) to be recognized as one consistent event. The representation must be coordinated between the two: In the previous example, if there is one highlighting, then there should also be one corresponding beep. When inconsistent, the interpretation of the feedback can be confusing, or only the dominant modality will be recognized.

3.3 Human Body Ergonomics (Motor Capabilities)

So far, we have mostly talked about human cognitive and perceptual capabilities and how display or input systems must be configured to match them. In this section, we briefly look at ergonomics aspects. To be precise, *ergonomics* is a discipline focused on making products and interfaces comfortable and efficient. Thus, broadly speaking, it encompasses mental and perceptual issues, although in this book, we restrict the term to mean ways to design interfaces or interaction devices for comfort and high performance according to the physical mechanics of the human body. For HCI, we focus on the human motor capabilities that are used to make input interaction. We start with Fitts's law and human motor control.

3.3.1 Fitts's Law

Fitts's law [23] is a model of human movement that predicts the time required to rapidly move to a target area as a function of the distance to and the size of the target. The movement task's Index of Difficulty (ID) can be quantified in terms of the required information amount, i.e., in the number of bits. From the main equation in Figure 3.19, the actual time to complete the movement task is predicted using a simple linear equation, where movement time, MT, is a linear function of ID.

$$MT = a + b * ID \text{ and } ID = \log(A/W + 1)$$

where A and B are coefficients specific to a given task.

Thus, to reiterate, ID represents an abstract notion of difficulty of the task, while MT is an actual prediction value for a particular task. The values for coefficients a and b are obtained by taking samples of the performance and mathematically deriving them by regression (Figure 3.20).

Note that the original Fitts's law was created for interaction with everyday objects (in the context of operation in factory assembly lines) rather than for computer interfaces. Researchers have applied the concept of Fitts's law to computer interfaces and have found that the same principle applies. For instance, as shown in Figure 3.21, the task of "dragging an icon into a trashcan icon" using a mouse can be assessed using Fitts's law [25]. Many other computer interactive tasks can be

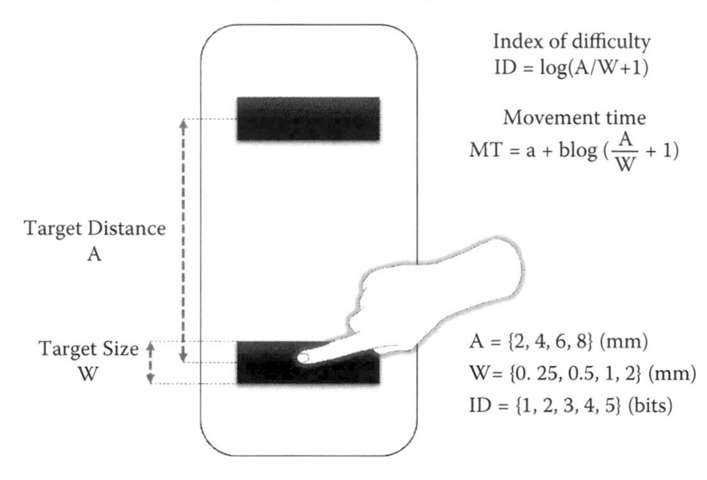

Figure 3.19 Illustration of Fitts's law. (From MacKenzie, I. S., *Human–Computer Interaction*, 7(1), 91–139, 1992 [24].)

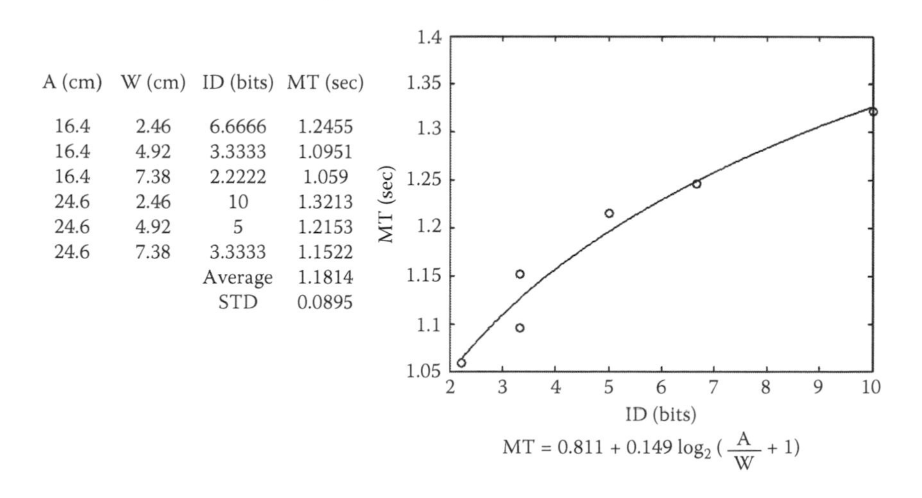

A (cm)	W (cm)	ID (bits)	MT (sec)
16.4	2.46	6.6666	1.2455
16.4	4.92	3.3333	1.0951
16.4	7.38	2.2222	1.059
24.6	2.46	10	1.3213
24.6	4.92	5	1.2153
24.6	7.38	3.3333	1.1522
		Average	1.1814
		STD	0.0895

Figure 3.20 Deriving the actual movement time by fitting based on samples of performance data. (From MacKenzie, I. S., *Human–Computer Interaction*, 7(1), 91–139, 1992 [24].)

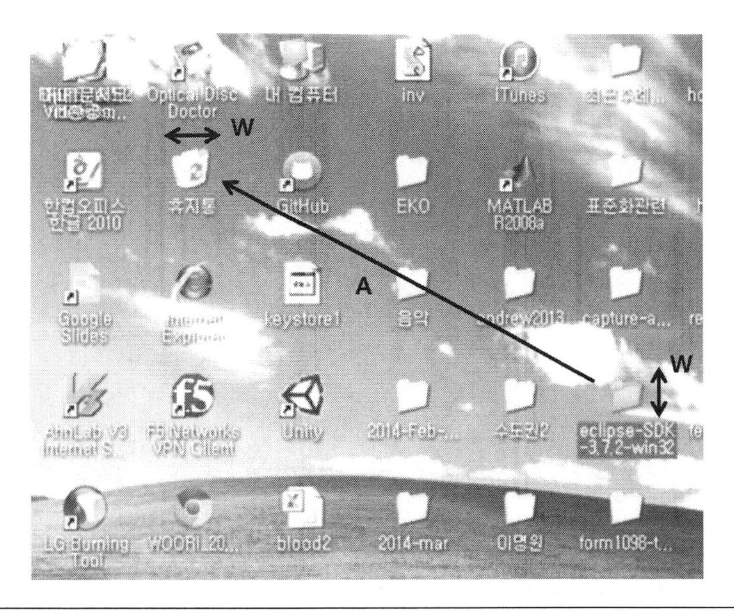

Figure 3.21 Applying Fitts's law to a computer interface (dragging a file icon into the trash-can icon). (From MacKenzie, I. S., Movement Time Prediction in Human–Computer Interfaces, in *Proceedings of the Conference on Graphics Interface '92*, Morgan Kaufman, San Francisco, 1992, pp. 140–150 [25].)

modeled similarly, and several revised Fitts's laws (e.g., for desktop computer interface, mobile interface) have been derived as well [24].

3.3.2 Motor Control

Perhaps the most prevalent form of input is made by the movements of our arms, hands, and fingers for keyboard and mouse input. Berard et al. have reported that there was a significant drop in human motor control performance below a certain spatial-resolution threshold [18]. For instance, while the actual performance is dependent on the form factor of the device used and the mode of operation, the mouse is operable with a spatial resolution on the order of thousands of dpi (dots per inch) or ≈0.020 mm, while the resolution for a 3-D stylus in the hundreds.

In addition to discrete-event input methods (e.g., buttons), modern user interfaces make heavy use of continuous input methods in the two-dimensional (2-D) space (e.g., mouse, touch screen) and increasingly in the 3-D space (e.g., haptic, Wii-mote). While the human capabilities will determine the achievable accuracy in such input methods, the control-display (C/D) ratio is often adjusted. C/D ratio refers to the ratio of the movement in the control device (e.g., mouse) to that in the display (e.g., cursor). If the C/D ratio is low, the sensitivity of the control is high and, therefore, travel time across the display will be fast. If the C/D/ ratio is high, sensitivity is low and, therefore, the fine-adjust time will be relatively fast.

Obviously, humans will exhibit different motor-control performances with different devices, as already demonstrated with the two types of device mentioned previously (e.g., mouse vs. 3-D stylus). The mouse and 3-D stylus, for instance, belong to what is called the *isometric* devices, where the movement of the device directly translates to the movement in the display (or virtual space). Nonisometric devices are those that control the movement in the display in principle with something else such as force, thus possibly with no movement input at all.

Control accuracy for touch interface presents a different problem. Despite our fine motor-control capability of submillimeter performance—and with recent touch screens offering higher than 4096-dpi resolution—it is the size of the fingertip contact (unless using a stylus pen), 0.3–0.7 cm, that makes it hard to make selection for relatively small objects. Even larger objects, once selected, are not easy to control if the touch screen is held by another hand or arm (i.e., unstable).

3.4 Others

There are many cognitive, perceptual, and ergonomic issues that have been left out. Due to the limited scope of this book, we only identify some of the issues for the reader to investigate further:

- Learning and adaptation
- Modalities other than the "big three" (visual/aural/haptic-tactile), such as gestures, facial expression, brain waves, physiological signals (electromyogram, heart rate, skin conductance), gaze, etc.
- Aesthetics and emotion
- Multitasking

3.5 Summary

In this chapter, we have reviewed the essence of human factors, including sensation, perception, information processing, and Fitts's law, as the foremost underlying theory for the design of interfaces for human–computer interaction. By the very principle of "Know thy user," it is clear that the HCI designer must have a basic understanding of these areas so that any interface will suit the user's most basic mental, perceptual, and ergonomic capabilities. We can also readily see that many of the HCI principles discussed previously in this book naturally derive from these underlying theories.

References

1. Norman, Donald A., and Stephen W. Draper. 1986. *User centered system design: New perspectives on human-computer interaction*. Boca Raton, FL: CRC Press.
2. Miller, George A. 1956. The magical number seven, plus or minus two: Some limits on our capacity for processing information. *Psychological Review* 63 (2): 81.
3. Marois, Rene, and Jason Ivanoff. 2005. Capacity limits of information processing in the brain. Trends in cognitive sciences 9 (6): 296–305.
4. Anderson, J. R., D. Bothell, M. D. Byrne, S. Douglass, C. Lebiere, and Y. Oin. 2004. An integrated theory of the mind. *Psychological Review* 111 (4): 1036–60.
5. Polk, T. A., and C. M. Seifert. 2002. *Cognitive modeling*. Cambridge, MA: MIT Press.
6. Salvucci, D. D., and N. A. Taatgen. 2008. Threaded cognition: An integrated theory of concurrent multitasking. *Psychological Review* 130 (1): 101–30.
7. Card, Stuart K., Thomas P. Moran, and Allen Newell. 1986. The model human processor: An engineering model of human performance. In *Handbook of human perception*. Vol. 2, *Cognitive processes and performance*, ed. K. R. Boff, L. Kauffman, and J. P. Thomas, 1–35. New York: John Wiley and Sons.
8. Schulz, Trenton. 2008. Using the keystroke-level model to evaluate mobile phones. In *Public systems in the future: Possibilities, challenges, and pitfalls*, Proceedings of the 31st Information Systems Research Seminar (IRIS31). Åre, Sweden.
9. Microsoft Research. 2013. CHI 2013: An immersive event (Illusions create an immersive experience). http://research.microsoft.com/en-us/news/features/chi2013-042913.aspx.
10. Ni, Tao, Greg S. Schmidt, Oliver G. Staadt, Mark A. Livingston, Robert Ball, and Richard May. 2006. A survey of large high-resolution display technologies, techniques, and applications. In *Proceedings of IEEE Virtual Reality Conference*, 223–36. Piscataway, NJ: IEEE.

11. Hemer, Mark A., Yalin Fan, Nobuhito Mori, Alvaro Semedo, and Xiaolan L. Wang. 2013. Projected changes in wave climate from a multi-model ensemble mark, *Nature Climate Change* 3:471–76.
12. Ware, C. 2012. *Information visualization: Perception for Design*. 3rd ed. Waltham, MA: Morgan Kaufmann.
13. Olson, Harry Ferdinand. 1967. *Music, physics and engineering*. Mineola, NY: Dover Publications.
14. Bregman, Albert S. 1994. *Auditory scene analysis: The perceptual organization of sound*. Cambridge, MA: MIT Press, A Bradford Book.
15. Apple. 2014. iOS7. http://www.apple.com/ios/siri.
16. Ferrucci, David. 2010. Building Watson. IBM Research. http://www.whitehouse.gov/sites/default/files/ibm_watson.pdf.
17. Wikipedia. 2014. Haptics. http://en.wikipedia.org/wiki/Haptic.
18. Bérard, François, Guangyu Wang, and Jeremy R. Cooperstock. 2011. On the limits of the human motor control precision: The search for a device's human resolution. In *Human-Computer Interaction–INTERACT*, 107–22. Berlin/Heidelberg: Springer.
19. Patel Prachi. 2010. Synthetic skin sensitive to the lightest touch. http://spectrum.ieee.org/biomedical/bionics/synthetic-skin-sensitive-to-the-lightest-touch.
20. Jones, Lynette A. 2000. Kinesthetic sensing. In *Proceedings of Workshop on Human and Machine Haptics*, 1–10. Cambridge, MA: MIT Press.
21. KU Leuven. 2010. Tactile feedback. https://www.mech.kuleuven.be/en/pma/research/ras/researchtopics/tactfb.html.
22. Reeves, L. M., J. Lai, J. A. Larson, S. Oviatt, T. S. Balaji, S. Buisine, P. Collings, et al. 2004. Guidelines for multimodal user interface design. *Communications of the ACM* 47 (1): 57–59.
23. Fitts, Paul M. 1954. The information capacity of the human motor system in controlling the amplitude of movement. *Journal of Experimental Psychology* 47 (6): 381.
24. MacKenzie, I. Scott. 1992. Movement time prediction in human-computer interfaces. In *Proceedings of the Conference on Graphics Interface '92*, 140–50. San Francisco: Morgan Kaufman.
25. MacKenzie, I. Scott. 1992. Fitts' law as a research and design tool in human-computer interaction. *Human-Computer Interaction* 7 (1): 91–139.

4
HCI Design

4.1 The Overall Design Process

In the first three chapters, we have studied notable principles, guidelines, and theories for the design of interfaces for human–computer interaction (HCI). In this book, HCI design is an integral part of a larger software design (and its architectural development) and is defined as the process of establishing the basic framework for user interaction (UI), which includes the following iterative steps and activities. HCI design includes all of the preparatory activities required to develop an interactive software product that will provide a high level of usability and a good user experience when it is actually implemented. We illustrate these four iterative steps using a concrete example after a short explanation of the respective steps (Figure 4.1).

- *Requirements analysis*: Any software design starts with a careful analysis of the functional requirements. For interactive software with a focus on the user experience, we take a particular look at functions that are to be activated directly by the user through interaction (*functional-task requirements*) and functions that are important in realizing certain aspects of the user experience (*functional-UI requirements*), even though these may not be directly activated by the user. One such example is an automatic functional feature of adjusting the display resolution of a streamed video based on the network traffic. It is not always possible to computationally separate major functions from those for the user interface. That is, certain functions actually have direct UI objectives. Finally, we identify *nonfunctional UI requirements*, which are UI features (rather than computational functions) that are not directly related to accomplishing the main application task.

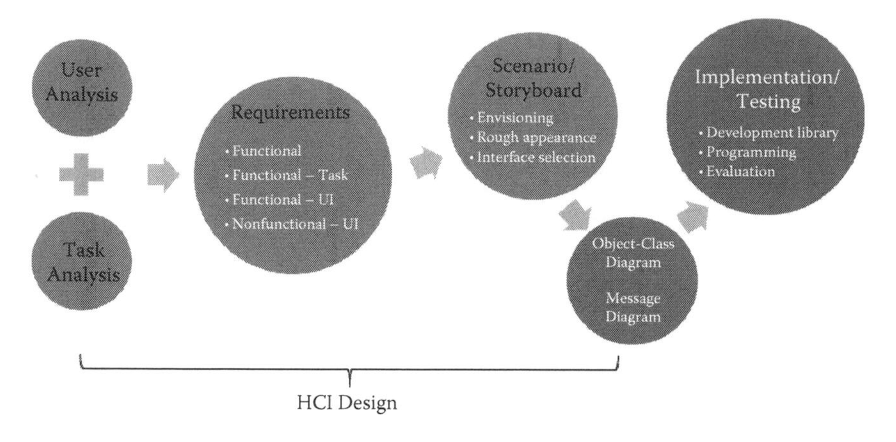

Figure 4.1 The overall iterative HCI design process (as a precursor to implementation).

For instance, requiring a certain font size or type according to a corporate guideline may not be a critical functional requirement, but a purely HCI requirement feature.

- *User analysis*: As we have emphasized previously, a user analysis is an essential step in HCI design. The results of the user analysis will be reflected back to the requirements, and this could identify additional UI requirements (functional or nonfunctional). It is simply a process to reinforce the original requirements analysis to further accommodate the potential users in a more complete way. For instance, a particular age group might necessitate certain interaction features such as a large font size and high contrast, or there might be need for a functional UI feature to adjust the scrolling speed.
- *Scenario and task modeling*: Equally important to user analysis is task analysis and modeling. This is the crux of interaction modeling: identifying the application task structure and the sequential relationships between the different elements. With a crude task model, we can also start to draw a more detailed scenario or storyboard to envision how the system would be used and to assess both the appropriateness of the task model and the feasibility of the given requirements. Again, one can regard this simply as an iterative process to refine the original rough requirements. Through the process of storyboarding, a rough visual profile of the interface can be sketched. Furthermore, the storyboard will serve as another

helpful medium in selecting the actual software or hardware interface. It will also serve as a starting point for drawing the object-class diagram, message diagrams, and the use cases for preliminary implementation and programming.

- *Interface selection and consolidation*: For each of the subtasks and scenes in the storyboard—particularly software interface components (e.g., widgets), interaction technique (e.g., voice recognition), and hardware (sensors, actuators, buttons, display, etc.)—choices will be made. The chosen individual interface components need to be consolidated into a practical package, because not all of these interface components may be available on a working platform (e.g., Android™-based smartphone, desktop PC, mp3 player). Certain choices will have to be retracted in the interest of employing a particular interaction platform. For instance, for a particular subtask and application context, the designer might have chosen voice recognition to be the most fitting interaction technique. However, if the required platform does not support a voice sensor or network access to the remote recognition server, an alternative will have to be devised. Such concessions can be made for many reasons besides platform requirements, such as due to constraints in budget, time, personnel, etc.

Before we go through a concrete example of HCI design, we first review possible and representative interfaces (hardware and software) to choose from in the following section.

4.2 Interface Selection Options

4.2.1 Hardware Platforms

Different interactions and subtasks may require various individual devices (sensors and displays). We take a look at the hardware options in terms of the larger computing platforms, which are composed of the usual devices. The choice of a design configuration for the hardware interaction platform is largely determined by the characteristics of the task/application that necessitates a certain operating environment. Therefore, the different platforms listed here are suited for and reflect various operating environments:

Figure 4.2 A basic desktop operating platform.

- *Desktop (stationary)*: Monitor (typical size: 17–42 in.; resolution: 1280 × 1012 or higher); keyboard, mouse, speakers/headphones (microphone) (Figure 4.2)
  Suited for: Office-related tasks, time-consuming/serious tasks, multitasking
- *Smartphones/handhelds (mobile)*: LCD screen (typical size: 3.5–5 in., resolution: 720 × 1280 or higher, weight ≈ 120 g), buttons, touch screen, speaker/headphones, microphone, camera, sensors (acceleration, tilt, light, gyro, proximity, compass, barometer), vibrators, mini "qwerty" keyboard
  Suited for: Simple and short tasks, special-purpose tasks
- *Tablet/pads (mobile)*: LCD screen (typical size: 7–10 in., resolution: 720 × 1280 or higher, weight ≈ 700 g), buttons, touch screen, speaker/headphones, microphone, camera, vibrators, sensors (acceleration, tilt, light, gyro, proximity, compass, barometer)
  Suited for: Simple, mobile, and short tasks, but those that require a relatively large screen (e.g., a sales pitch; see Figure 4.3)
- *Embedded (stationary/mobile)*: LCD/LED screen (typical size: less than 3–5 in., resolution: low), buttons, special sensors, and output devices (touch screen, speaker, microphone, special sensors); embedded devices may be mobile or stationary and offer only very limited interaction for a few simple functionalities (Figure 4.4)

Figure 4.3 Making sales with padlike devices leveraging on their mobility and relatively large screen size.

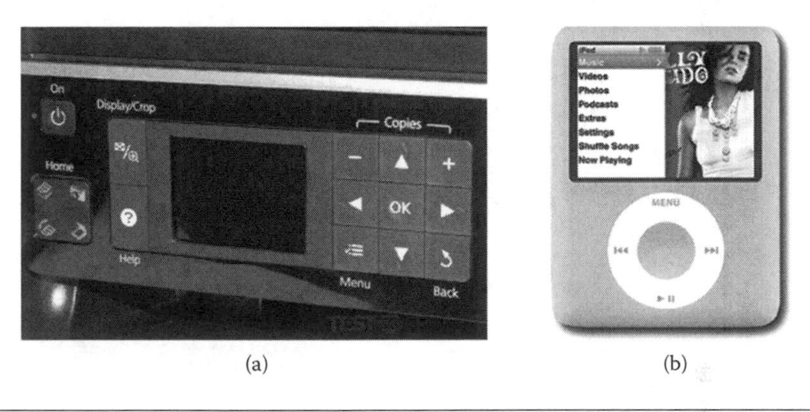

(a) (b)

Figure 4.4 Embedded interaction platform: (a) mobile and (b) stationary.

Suited for: Special tasks and situations where interaction and computations are needed on the spot (e.g., printer, rice cooker, mp3 player, personal media player)

- *TV/consoles (stationary)*: LCD/LED screen (typical size: >42 in., resolution: HD), button-based remote control, speaker, microphone, game controller, special sensors, peripherals (camera, wireless keyboard, Wii mote–like device [1], depth sensor such as Kinect [2, 3]) (Figure 4.5)

  Suited for: TV-centric tasks, limited interaction, tasks that need privacy (e.g., wild-gesture-based games in the living room)

- *Kiosks/installations (stationary)*: LCD screen (typical size: 10–13 in., resolution: low to medium), buttons, speaker, touch

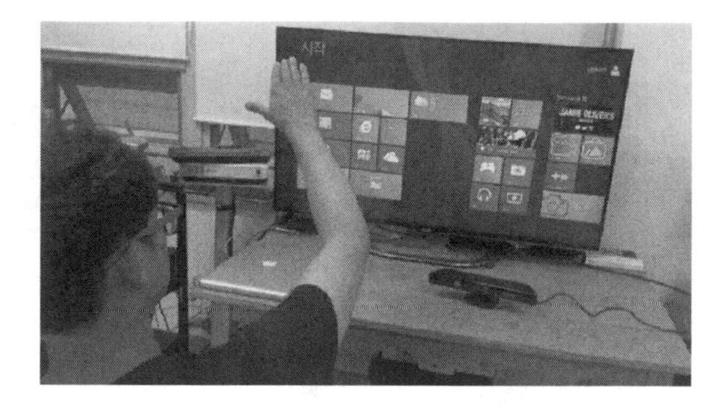

Figure 4.5 TV/console interaction environment (Xbox LIVE, http://www.xbox.com/ko-KR/Live/what-is-live?xr=shellnav).

screen, special sensors, peripherals (microphone, camera, RFID/credit-card reader, heavy-duty keyboard) (Figure 4.6)
Suited for: Public users and installations, limited interaction, short series of selection tasks, monitoring tasks

- *Virtual reality (stationary)*: Large-surround and high-resolution projection screen/head-mounted display/stereoscopic display, 3-D tracking sensors, 3-D sound system, haptic/tactile display, special sensors, peripherals (microphone, camera, depth sensors, glove) (Figure 4.7)
  Suited for: Spatial training, tele-experience and tele-presence, immersive entertainment
- *Free form (stationary and mobile)*: Special-purpose hardware platforms consisting of a customized configuration of individual devices best suited for a given task (when cost is not the

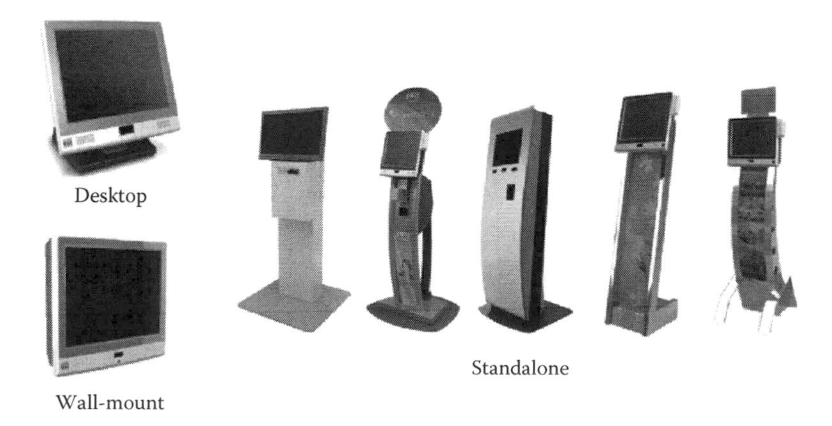

Figure 4.6 Various kiosk/installation types of interaction platform.

Figure 4.7 Virtual-reality interaction platform (Visbox, http://www.visbox.com/imgs/viscube-hd.html).

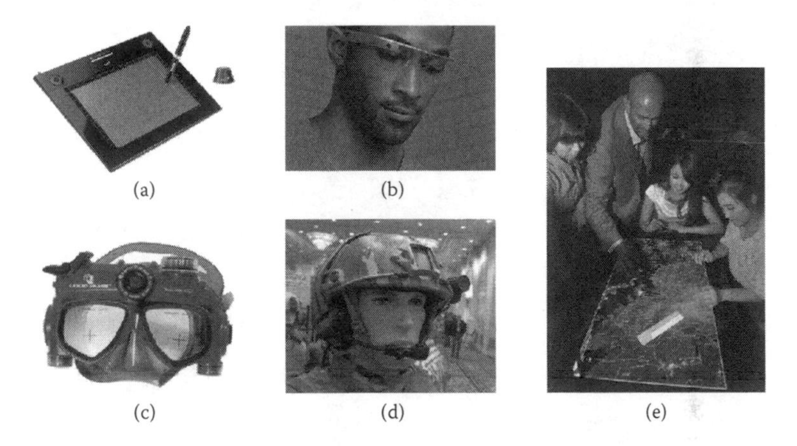

(a) (b) (c) (d) (e)

Figure 4.8 Examples of special-purpose interfaces (from top left, counterclockwise): (a) pen tablet (Genius, KYE Systems G-Pen, http://www.geniusnet.com/wSite/ct?xItem=16835&ctNode=174), (b) a glass-type see-through heads-up display (Google® Glass, http://www.google.com/glass), (c) camera-integrated scuba gear (Liquid Image Scuba, http://www.liquidimageco.com/collections/scuba), (d) special military helmet for tactical command and control (Defensereview, http://www.defensereview.com; photo credit: David Crane), (e) multitouch tabletop platform for multiple users (Samsung MultiTouch Display, http://www.samsung.com/sec/news/presskit/hf).

biggest factor). (There are many such custom-designed interfaces, such as those shown in Figure 4.8.)

4.2.2 Software Interface Components

Most of these software components are quite well known and familiar to most of the readers, so we only highlight important issues to consider in the interface selection.

- *Windows/layers*: Modern desktop computer interfaces are designed around *windows*, which are visual output channels and abstractions for individual computational processes. For a single application, a number of subtasks may be needed concurrently and thus must be interfaced through multiple windows. One window among the many (or task) would be "active," and this window becomes "focused" by placing the mouse cursor over it or by an explicit click. For relatively large displays, overlapping windows may be used. However, as the display size decreases (e.g., mobile devices), *nonoverlapping layers* (a full-screen window) may be used in which individual layers are activated in turn by "flipping" through them (e.g., flicking movements on touch screens) (Figure 4.9).

  While multiple overlapped windows have been traditionally used for relatively large desktop platforms and layers for smaller devices, with the recent trend and requirement of "multiple device and single-user experience," the Windows® Metro–style interface has unified the two [4]. Even on the desktop, the Metro style presents individual applications on the full screen without marked borders, but instead offers new convenient means for sharing data with other applications and switching between the applications or tasks (Figure 4.10). Other important detailed considerations for a window (for supporting interaction for a subtask) might be its size, interior layout, and management method (e.g., activation, deactivation, suspension).

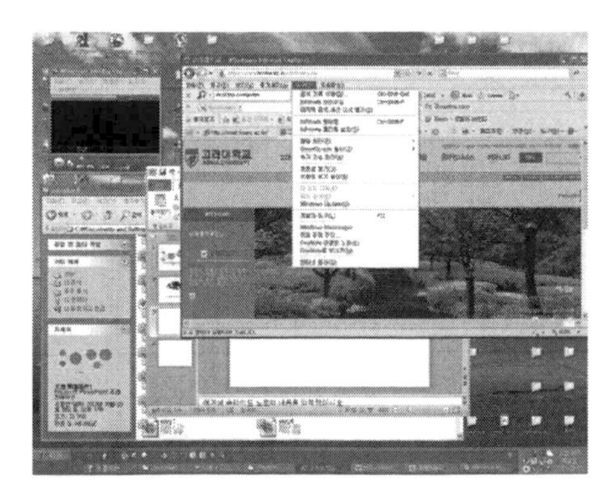

Figure 4.9 The hallmark of modern desktop user interfaces: multiple overlapping windows (left) and nonoverlapping layers for smaller displays (right).

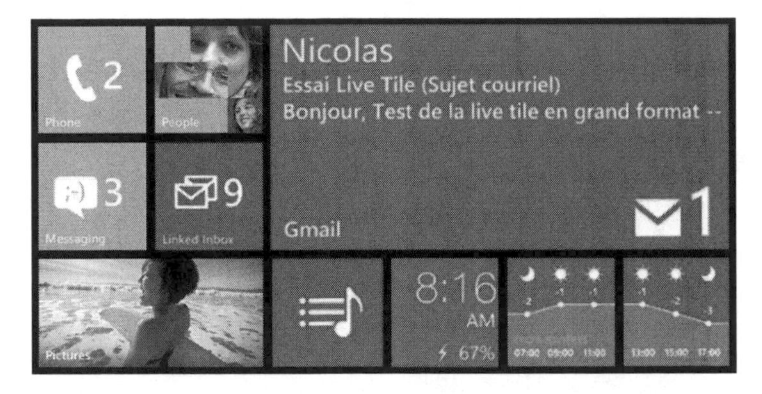

Figure 4.10 The Microsoft® Metro–style interface that unifies the mobile and desktop interaction.

- *Icons*: Interactable objects may be visually represented as a compact and small pictogram such as an icon (and similarly as an "earcon" for the aural modality). Clickable icons are simple and intuitive (Figure 4.11). As a compact representation designed for facilitated interaction, icons must be designed to be as informative or distinctive as possible despite their small size and compactness.

  The recent Windows Metro–style interface has introduced a new type of icon called a *tile* that can dynamically change its look with useful information associated with what the icon is supposed to represent [5]. For instance, the e-mail application icon dynamically shows the number of new unread e-mails (Figure 4.10).
- *Menus*: Menus allow activations of commands and tasks through selection (recognition) rather than recall. Typical menus are

Figure 4.11 (Left) Obscure Google Chrome browser icon design (difficult to tell at a glance what the icon represents) vs. (right) informative icon design for the Angry Birds application (the visual imagery of an "angry bird" is well captured).

organized as a one-dimensional list or a two-dimensional (2-D) array of items (represented in text or as icons/earcons).

Selection of a menu item involves three subtasks: (a) activating the menu and laying out the items (if not already activated by default), (b) visually scanning and moving through the items (and scrolling if the display space is not sufficient to contain and show the whole menu of items at once), and (c) choosing the wanted item. All of these subtasks are realized by making discrete inputs, e.g., by mouse click, screen touch, button push, voice command, etc.

Menus (i.e., lists of items) may be presented in a variety of styles and mechanisms, as seen in Figure 4.12. Some of the most popular ones are shown in Figure 4.13: the pull-down, pop-up, toolbars, tabs, scroll menu, 2-D array of icons, buttons, check boxes, hot keys, etc. Table 4.1 shows how to best use these different types of menus.

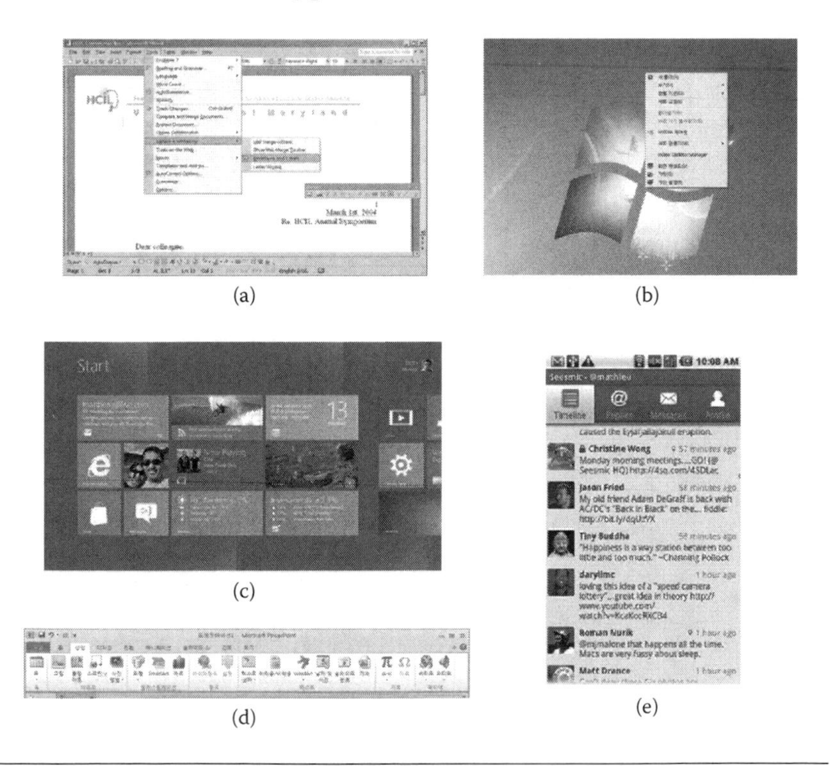

Figure 4.12 Different styles of menus 1: (a) pull down, (b) pop up, (c) 2-D application bar, (d) 1-D toolbar, and (e) tabs. (From Petzold, C., Microsoft XNA Framework Edition: Programming Windows Phone 7, Microsoft Press, 2010 [4].)

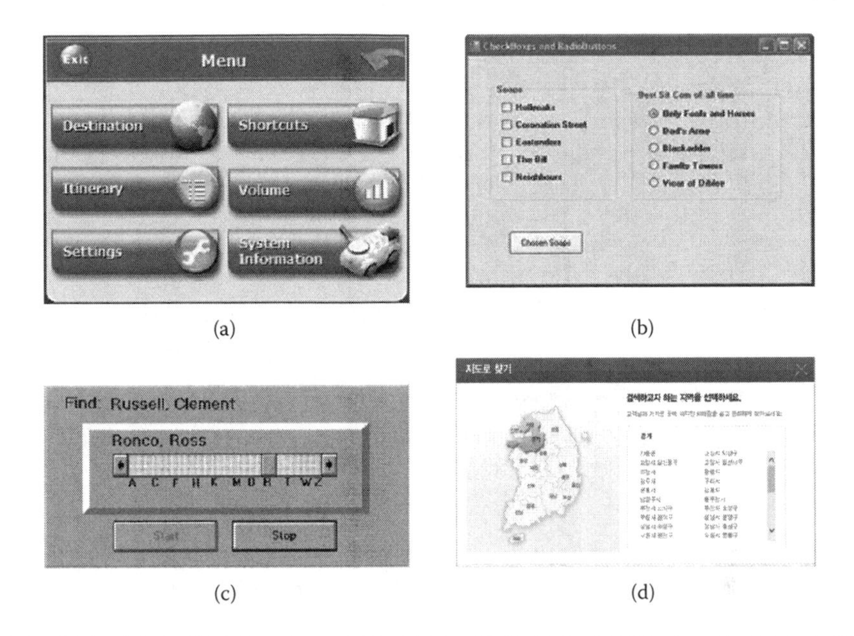

Figure 4.13 Different styles of menus 2: (a) buttons, (b) check boxes and radio buttons, (c) slider menu, (d) image map.

Table 4.1 Where to Use Different Menu Styles

MENU TYPE	USAGE
Pull down	Top level (main) categorical menu
Pop up	Object specific, context specific
Toolbar	Functional/operational tasks
Tabs	File folder metaphor (categorical menu)
Scroll menu	Long menu (many menu items)
2-D array/Image maps	Identification of items by icons (vs. by long names) or pictures
Buttons/Hyperlinks	Short menu (few choices)
Check boxes/Radio buttons	Multiple choice/exclusive choice
Hot keys	For expert users
Aural menu	Telemarketing and for use by the disabled

The menu items are usually subtasks to be invoked or the target interaction objects for certain tasks to be operated upon. In either case, it is clear that the menu must be organized, categorized, and structured (typically hierarchically) according to the task and the associated objects. At each level of the menu, the number of items should be managed to be ideally below the "magic number 8" (the limit of our short-term memory, discussed in Section 3.1.2) [6]. However, it may not always be possible to achieve such a design or model. If long menus are inescapable, the items should at least be laid out in a systematic manner, e.g., in the order of their frequency, importance, alphabetic sequence, etc.

- *Direct interaction*: The mouse/touch-based interaction is strongly tied to the concept of *direct and visual interaction*. Before the mouse era, the HCI was mostly in the form of keyboard inputting of text commands. The mouse made it possible for users to apply a direct metaphoric "touch" upon the target objects (which are visually and metaphorically represented as concrete objects with the icons) rather than "commanding" the operating system (via keyboard input) to indirectly invoke the job. In addition to this virtual "touch" for simple enactment, the direct and visual interaction has further extended to direct manipulation, e.g., moving and gesturing with the cursor against the target interaction objects. "Dragging and dropping," "cutting and pasting," and "rubber banding" are typical examples of these extensions.
- *GUI components*: Software interaction objects are mostly visual. We have already discussed the windows, icons, menus, and mouse/pointer-based interactions, which are the essential elements for the graphical user interface (GUI), also sometimes referred to as the WIMP (window, icon, mouse, and pointer) [7]. The term *WIMP* is deliberately chosen for its negative connotation to emphasize its contrast with a newer upcoming generation of user interfaces (such as voice/language and gesture based). However, WIMP interfaces have greatly contributed to the mass proliferation of computer technologies. In Chapter 5, we will take a more systematic look at the GUI components as part of implementation knowledge. For now, in considering interface options, it suffices to understand the following representative GUI components, aside from those for discrete selection (WIMP), for soliciting input from a user in a convenient way (Figure 4.14):
  - *Text box*: Used for making short/medium alphanumeric input
  - *Toolbar*: A small group of frequently used icons/functions organized horizontally or vertically for a quick direct access
  - *Forms*: Mixture of menus, buttons, and text boxes for long thematic input

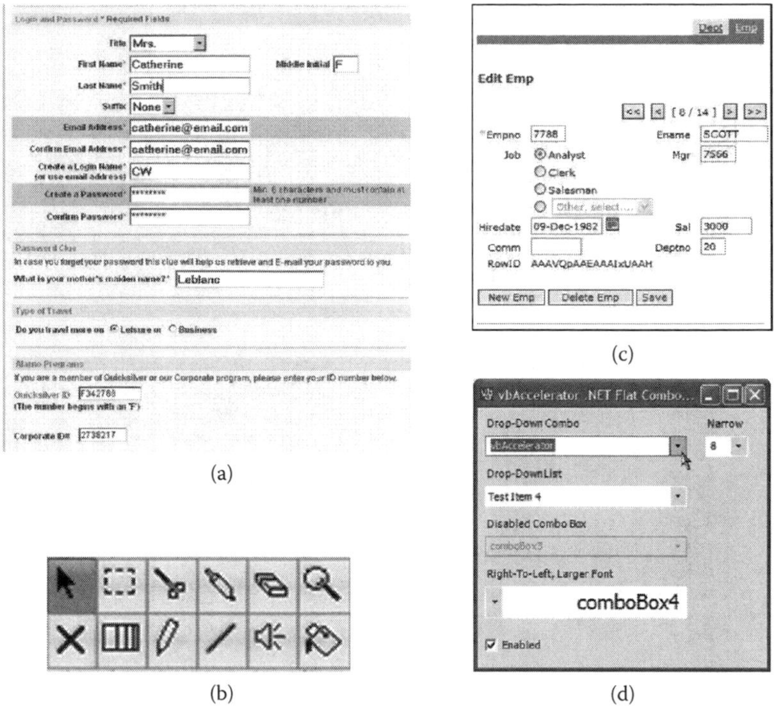

(a)

(b)

(c)

(d)

Figure 4.14 GUI interface components: (a) form, (b) toolbar, (c) dialog box, (d) combo box.

- *Dialog/combo boxes*: Mixture of menus, buttons, and text boxes for short mixed-mode input
- *3-D interface (in 2-D interaction input space)*: Standard GUI elements that are operated and presented in the 2-D space, i.e., they are controlled by a mouse or touch screen and laid out on a 2-D screen. However, 2-D control in a 3-D application is often not sufficient (e.g., 3-D games). The mismatch in the degrees of freedom brings about fatigue and inconvenience (Figure 4.15). For this reason, non-WIMP–based interfaces such as 3D motion gestures are gaining popularity.

  Aside from a task such as 3-D games and navigation, it is also possible to organize the 2-D-operated GUI elements in 3-D virtual space. It is not clear whether such an interface brings about any particular advantages because, despite the added dimension, the occlusion due to overlap will remain, as the interface is viewed from only one direction (into the screen). In fact, the user can be burdened with the added

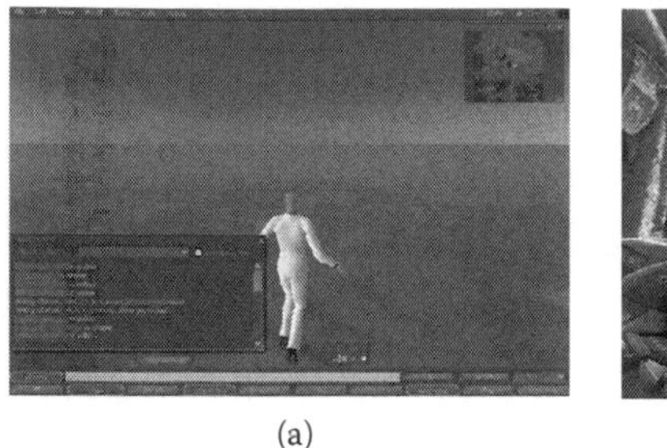

(a)

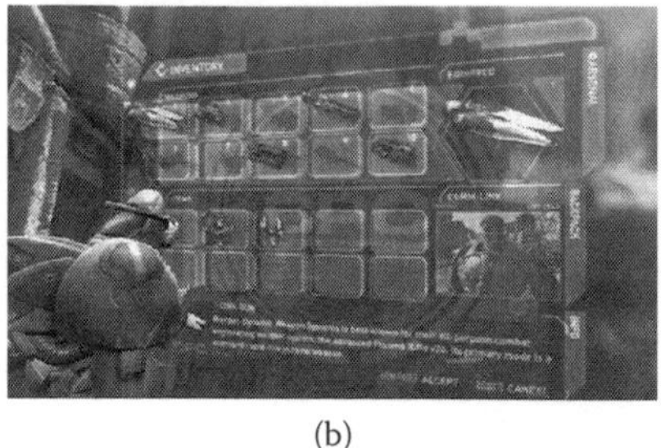

(b)

Figure 4.15 3-D interface in 2-D interaction input space (e.g., mouse) for (a) 3-D task such as spatial navigation (SecondLife, http://secondlife.wikia.com/wiki/User_Interface) and (b) 2-D GUI elements laid out in 3-D space (EpicGames Scaleform GFx, https://udn.epicgames.com/Three/Scaleform.html) (mainly for futuristic "wow" factor).

control if one needs to place or manipulate GUI objects in three dimensions. However, it is sometimes employed anyway in 3-D games for aesthetic reasons and the "wow" factor.

- *Other (non-WIMP) interfaces*: The WIMP interface is synonymous with the GUI. It has been a huge success since its introduction in the early 1980s, when it revolutionized computer operations. Thanks to continuing advances in interface technologies (e.g., voice recognition, language understanding, gesture recognition, 3-D tracking) and changes in the computing environment (e.g., personal to ubiquitous, sensors everywhere)—new interfaces are starting to making their way into our everyday lives. In addition, the cloud-computing environment has enabled running computationally expensive interface algorithms, which non-WIMP interfaces often require, over less powerful (e.g., mobile) devices against large service populations. Chapters 7–9 in this book take a look at some basic implementation issues for these new non-WIMP interfaces.

4.3 Wire-Framing

The interaction modeling and interface options can be put together concretely using the so-called wire-framing process. Wire-framing originated from making rough specifications for website page design and resembles scenarios or storyboards. Usually, wire-frames look like page schematics or screen blueprints, which serve as a visual guide

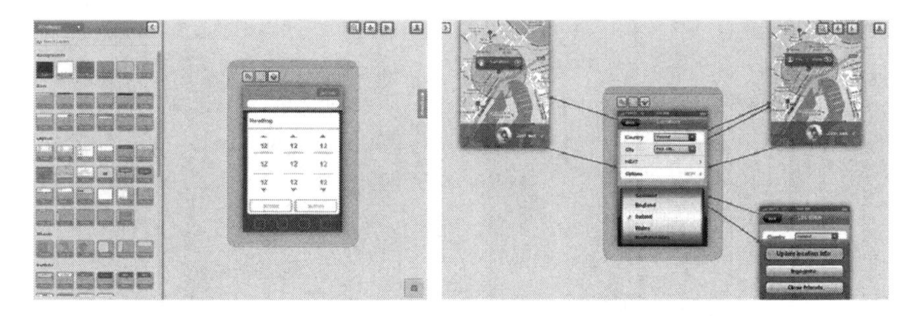

Figure 4.16 An example of a wire-framing tool (FluidUI, https://www.fluidui.com). Designing the content of a screen (left) and overall interaction behavior, e.g., how screens switch upon interaction (right).

that represents the skeletal framework of a website or interface [8]. It depicts the page layout or arrangement of the UI objects and how they respond to each other. Wireframes can be pencil drawings or sketches on a whiteboard, or they can be produced by means of a broad array of free or commercial software applications. Figure 4.16 shows such a wire-framing tool. Wireframes produced by these tools can be simulated to show interface behavior, and depending on the tools, the interface logic can be exported for actual code implementation (but usually not). Note that there are tools that allow the user to visually specify UI elements and their configuration and then automatically generate code. Regardless of which type of tool is used, it is important that the design and implementation stages be separated. Through wire-framing, the developer can specify and flesh out the kinds of information displayed, the range of functions available, and their priorities, alternatives, and interaction flow.

4.4 "Naïve" Design Example: No Sheets 1.0

4.4.1 Requirements Analysis

To illustrate the HCI design process more concretely, we will go through the design of a simple interactive smartphone (Android) application, called *No Sheets*. The main purpose of this application is to use the smartphone to present sheet music,* thereby eliminating the need to handle paper sheet music (Figure 4.17). An initial requirements list may look something like the one in Table 4.2. Note that this again

* Sheet music is a written recording of music which is transcribed in music notation.

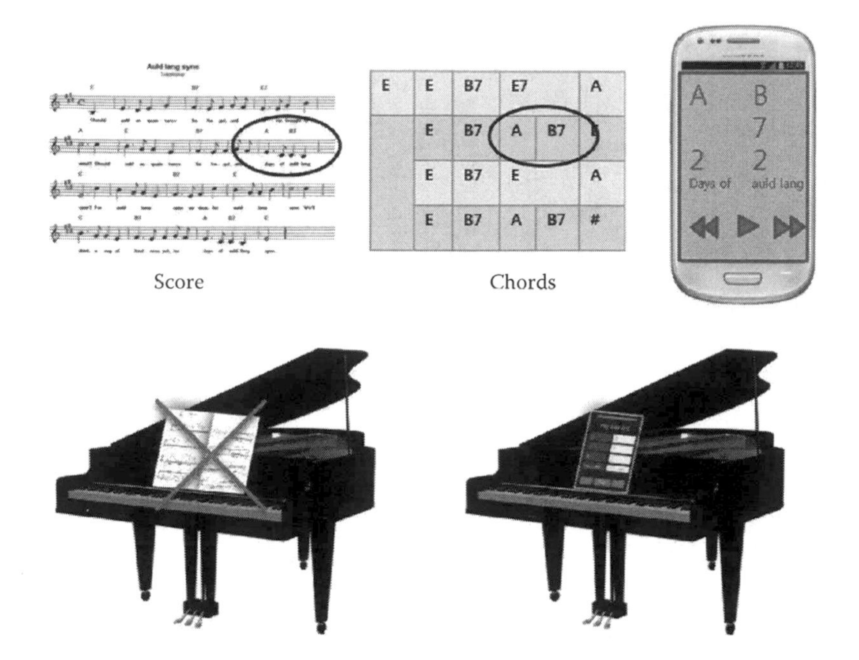

Figure 4.17 No Sheets: Replacing paper sheet music with the smartphone. No more flying pages; no more awkward flipping and page searching.

Table 4.2 Initial Requirements for No Sheets

1. Use the smartphone to present transcribed music like "sheet music." Transcription includes only those for basic accompaniment like the chord information (key and type such as C# dom7), beat information (e.g., second beat in the measure).
2. Eliminate the need to carry and manage physical sheet music. Store music transcription files using a simple file format.
3. Help the user effectively accompany the music by timed and effective presentation of musical information (e.g., paced according to a preset tempo).
4. Help the user effectively practice the accompaniment and sing along through flexible control (e.g., forward, review, home buttons).
5. Help user sing along by showing the lyrics and beats in a timed fashion.

would be part of any software development process. Here, we focus more on the HCI-related requirements for the sake of brevity.

4.4.2 User Analysis

The typical user for No Sheets is a smartphone owner and novice/intermediate piano player (perhaps someone who wants to show off their musical skill at a piano bar) (Figure 4.18). Since a smartphone is

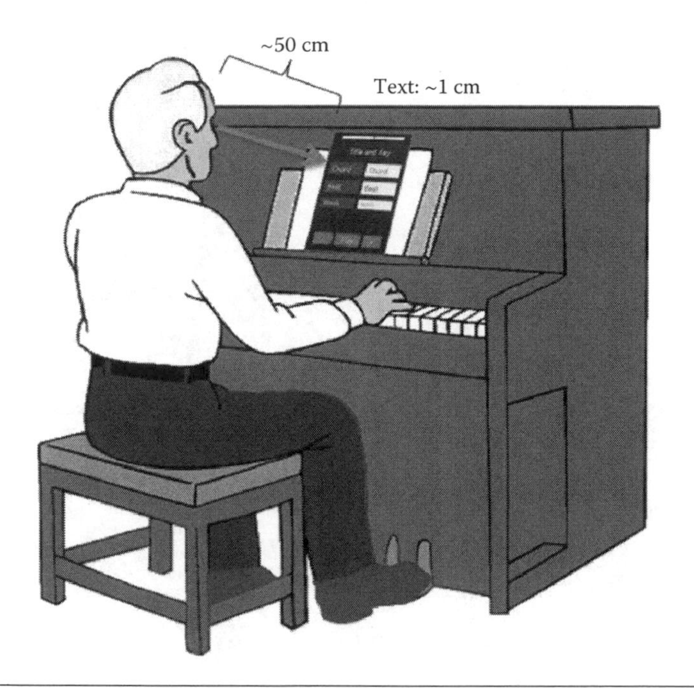

Figure 4.18 A typical usage situation for No Sheets.

used, we would have to expect a reasonable power of sight for a typical usage (e.g., a viewing distance of about 50 cm subtending a letter of ±1 cm). There does not seem to be a particular consideration for a particular age group or gender. However, there may be a consensus on how the chord/music information should be displayed (e.g., portrait vs. landscape, information layout and locations of the control buttons, color-coding method, up-down scrolling vs. left-right paging, etc.). A very minimal user analysis (that of the developer himself) resulted in (naïve first trial) interface requirements as shown in Table 4.3. Note that, for now, most of the requirements or choices are rather arbitrary without clear justifications. A revised design based on a more careful user analysis/evaluation is presented in Chapter 8.

4.4.3 Making a Scenario and Task Modeling

Based on the short requirements in Table 4.3, we derive a hierarchical simple task model as shown in the following list and in Figure 4.19. Each task is to be activated directly by the user through an interface.

- *Select song*: Select the song to view

Table 4.3 User Interface Requirements from a Very Minimal User Analysis

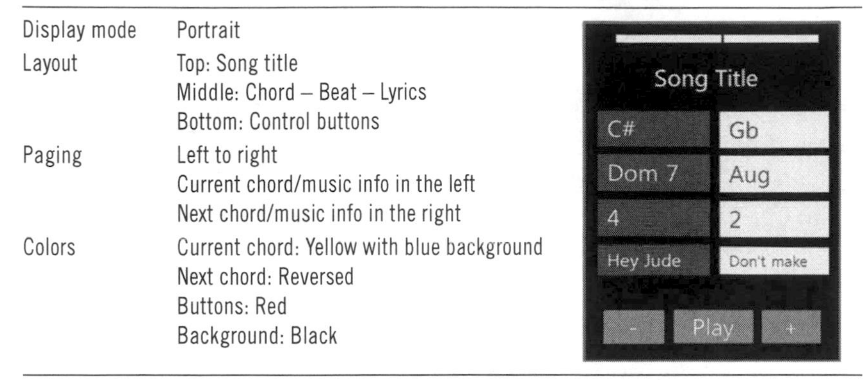

Display mode	Portrait
Layout	Top: Song title Middle: Chord – Beat – Lyrics Bottom: Control buttons
Paging	Left to right Current chord/music info in the left Next chord/music info in the right
Colors	Current chord: Yellow with blue background Next chord: Reversed Buttons: Red Background: Black

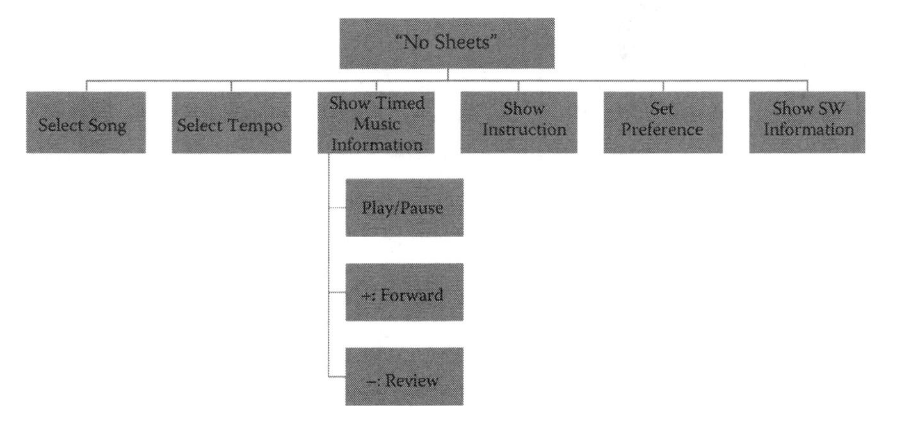

Figure 4.19 A simple task model for No Sheets. The top-level application has six subtasks (select song, select tempo, etc.), and the third subtask (show music info) has yet other subtasks: play/pause, fast-forward, and review.

- *Select tempo*: Set the tempo of the paging
- *Show timed music information*: Show the current/next chord/beat/lyric
  - Play/Pause: Activate/deactivate the paging
  - Fast-forward: Manually move forward to a particular point in the song
  - Review: Manually move backward to a particular point in the song
- *Show instruction*: Show the instruction as to how to use the system
- *Set preferences*: Set preferences for information display and others

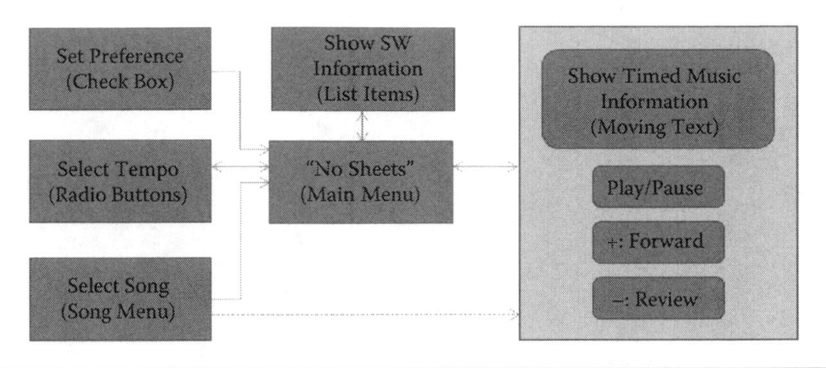

Figure 4.20 A possible state-transition diagram for No Sheets.

- *Show software information*: Show version number and developer information

The subtasks, as actions to be taken by the user, can be viewed computationally as action events or, reversely, as states that are activated according to the action events. Figure 4.20 shows a possible state-transition diagram for No Sheets. Through such a perspective, one can identify the precedence relationship among the subtasks. From the top main menu (middle of the figure), the user is allowed to set/select/change/view the preference, tempo, song, and software information. The user is also able to play and view the timed display of the musical information, but only after a song has been chosen (indicated by the dashed arrow). While the timed music information is displayed, the user can concurrently—the four states (or equivalently actions) in the transparent box in the right are concurrent—play/stop, move forward, and move backward. Such a model can serve as a rough starting point for defining the overall software architecture for No Sheets.

A storyboard is then drawn based on the task model to further envision its usage and possible interface choices. (A storyboard consists of graphic illustrations organized in sequence and is often used to previsualize motion picture, animation, and interactive experiences.) There is no fixed format, but each illustration usually includes a depiction of the important steps in the interaction, annotated with a description of important aspects (e.g., possible interface choice, operational constraints and any special consideration needed, and usage contexts). Figures 4.21–4.25 show the initial storyboards for No Sheets, illustrating the motivational context, a typical usage scenario and sequences, and rough mobile interface sketches.

Figure 4.21 Motivational context for No Sheets.

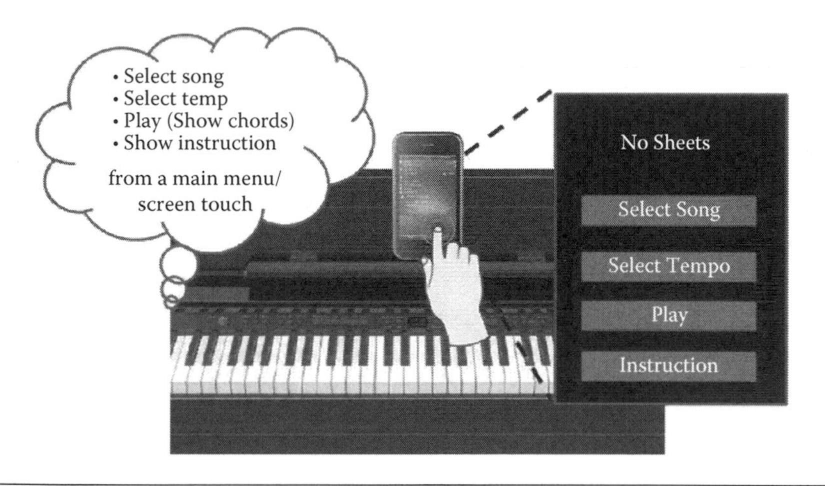

Figure 4.22 A typical usage scene 1: The top-level menu.

4.4.4 Interface Selection and Consolidation

We end this exercise by finalizing the choice of particular interfaces for the individual subtasks. Table 4.4 shows the final decision and justifications. It is very important that we try to adhere to the HCI principles, guidelines, and theories to justify and prioritize our decision. Note that we have started with the requirement that the application is to be deployed on a smartphone (the interface platform). Again, our

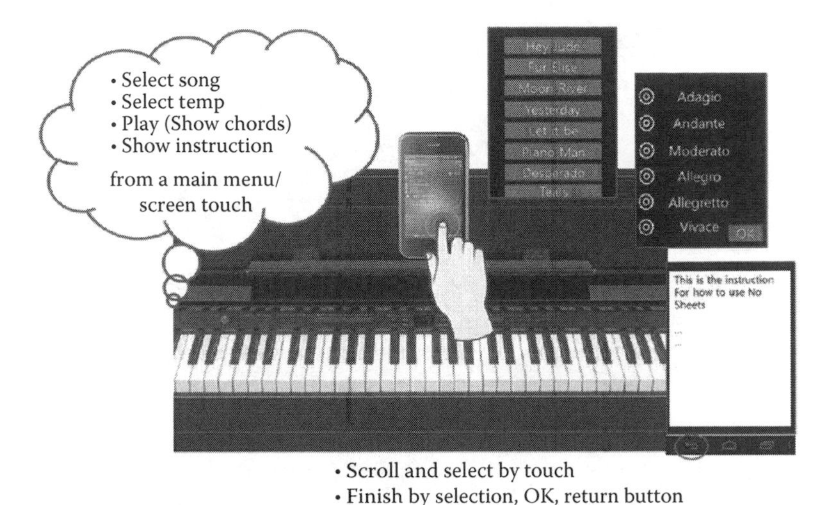

Figure 4.23 A typical usage scene 2: Interface looks for three subtasks (e.g., song selection, tempo selection, and showing instruction).

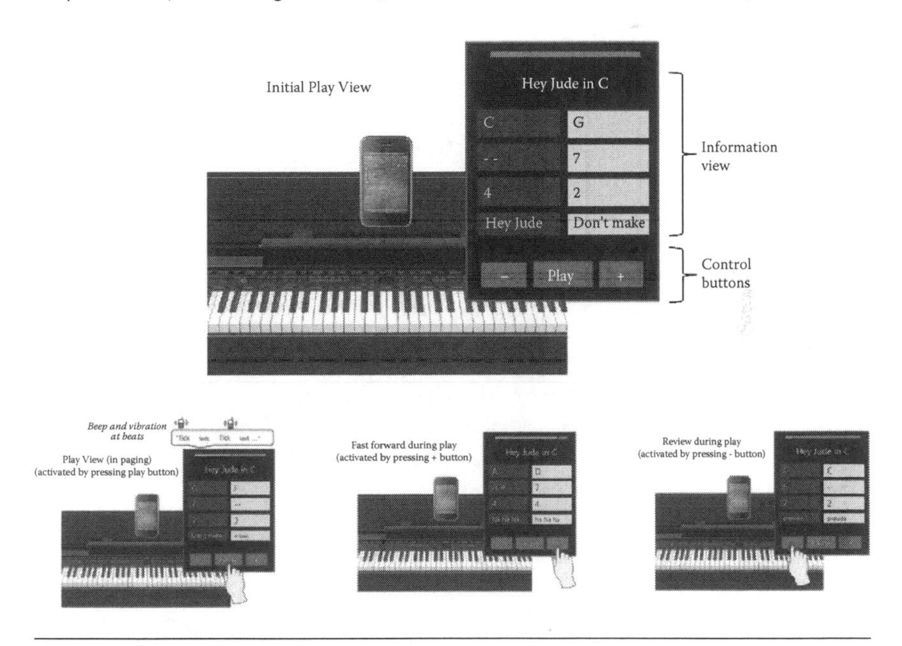

Figure 4.24 A typical usage scene 3: Interface looks during "play" and the three concurrently activatable subtasks (play/pause, move forward, and move backward).

initial choice will purposely not be well thought out, just to illustrate that a naïve and hurried choice would expose the application to be at high risk of eventually failing in terms of usability and user experience, even if it computationally satisfies the required functionalities.

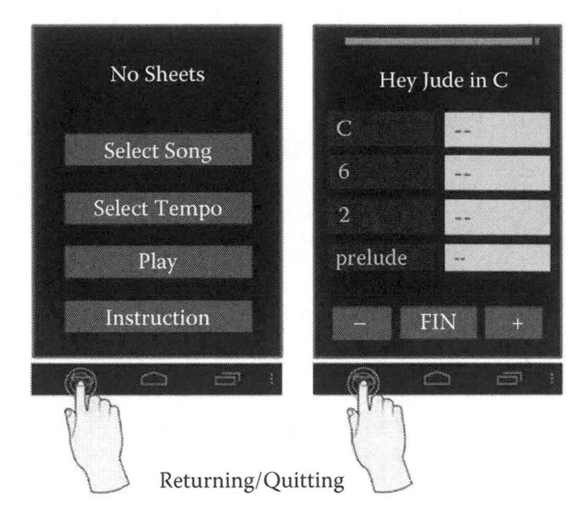

Figure 4.25 A typical usage scene 4: Moving between views/stages and quitting the application by using the standard Android menu button interface.

Table 4.4 Initial Finalization of the Interface Design Choice for No Sheets

SUBTASK	INTERFACE DESIGN CHOICE	JUSTIFICATION
Invoking main functions	• Touch menu • Menu items in red	• Familiar interface • Catch attention
Selecting/changing song	• Scrolling menu • Return to main menu upon selection	• There may be many songs
Selecting/changing tempo	• Scrolling radio buttons • Return to main menu by OK button	• Only one tempo is chosen at a given time
Showing instruction	• Show a one page/screen image with condensed instructional content	• Present condensed content
Playing/pause (view)	• Show progress bar on top • Control interface in the bottom • Provide sound beeps and vibration for first and second beat • Color-code different types of information	• Show status • Familiar interface • Use multimodal feedback for redundancy
Moving forward (+)	• Forward button on the right	• Cultural consideration (moving from left to right) • Show status through progress bar
Moving backward (–)	• Backward button on the left	• Cultural consideration (moving from left to right) • Show status through progress bar
Quitting	• Use platform button	• Use platform (e.g., Android) guideline

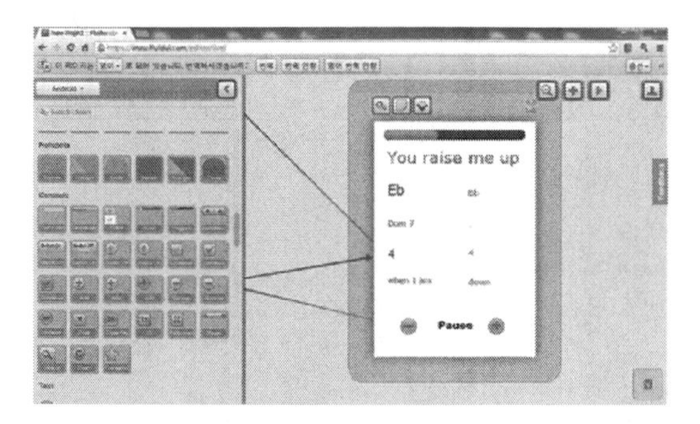

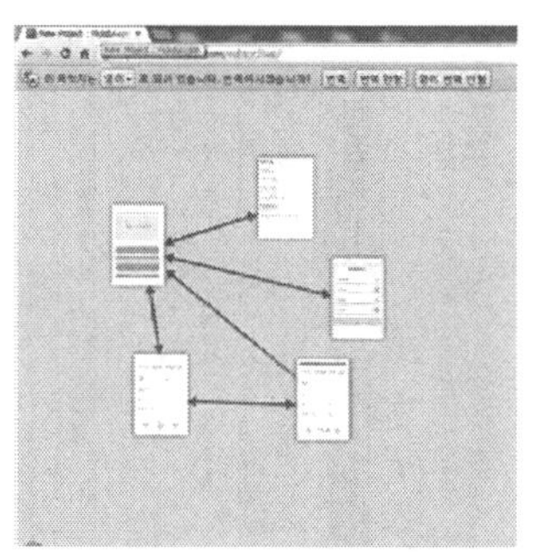

Figure 4.26 Initial design wireframe for No Sheets 1.0 using a wire-framing tool. Left: Icons and GUI elements in the menu in the left can be dragged onto the right to design the interface layer. Right: Navigation among the design layers can be defined as well (indicated by the arrows).

This will become more apparent as we evaluate the initial prototype and revise our requirements and design for No Sheets 2.0 (presented in Chapter 8). Figure 4.26, based on the use of a commercial wire-framing tool, shows the final interface look and the interaction flow.

4.5 Summary

In this chapter, we have described the design process for interactive applications, focusing on modeling of the interaction and selection of the interface. The discussion started with a requirements analysis and its continued refinement through user research and application-task modeling. Then, we drew up a storyboard and carefully considered different options for particular interfaces by applying any relevant HCI principles, guidelines, and theories. The overall process was illustrated with a specific example design process for a simple application. It roughly followed the aforementioned process, but it did so (purposefully) in a hurried and simplistic fashion, leaving much potential for later improvement. Nevertheless, this exercise emphasizes that the design process is going to be unavoidably iterative, because it is not usually possible to have the provisions for all usage possibilities. This is why an evaluation is another necessary step in a sound HCI design cycle, even if a significant effort is thought to have gone into the initial design and prototyping. In the next chapters, we first look at issues involved with taking the design into actual implementation. The implemented prototype (or final version) must then be evaluated in real situations for future continued iterative improvement, extension, and refinement.

References

1. Nintendo. 2013. Wii. http://web.archive.org/web/20080212080618/http://wii.nintendo.com/controller.jsp.
2. Freedman, Barak, Alexander Shpunt, Meir Machline, and Yoel Arieli. 2010. Depth mapping using projected patterns. U.S. Patent 12/522,171. Filed Apr. 2, 2008, and issued May 13, 2010.
3. Microsoft. 2014. Kinect for XBox. http://www.xbox.com/en-us/kinect/.
4. Petzold, Charles. 2010. *Programming Windows Phone 7: Microsoft XNA framework edition*. Redmond, WA: Microsoft Press.
5. Freeman, Adam. 2012. *Metro revealed: Building Windows 8 apps with HTML5 and JavaScript*. New York: Apress Media.
6. Miller, George A. 1956. The magical number seven, plus or minus two: Some limits on our capacity for processing information. *Psychological Review* 63 (2): 81.
7. Van Dam, Andries. 1997. Post-WIMP user interfaces. *Communications of the ACM* 40 (2): 63–67.
8. Brown, Dan M. 2010. *Communicating design: Developing web site documentation for design and planning*. Berkeley, CA: New Riders.

5
User Interface Layer

5.1 Understanding the UI Layer and Its Execution Framework

Interactive applications are implemented and executed using the user interface (UI) software layers (collectively the UI layer). The UI layer refers to a set of software that operates above the core operating system (and underneath the application). It encapsulates and exposes system functions for

- Fluid input and output (I/O)
- Facilitation of development of I/O functionalities (in the form of an application programming interface/library [API] or toolkit)
- Run-time management of graphical applications and UI elements often manifested as windows or graphical user interface (GUI) elements (in the form of separate application often called the window manager)

Since most interfaces are graphical, the UI layer uses a two- or three-dimensional (2-D or 3-D) graphical system based on which GUI elements are implemented (lower part of Figure 5.1). Thus, to summarize, the UI layer is largely composed of (a) an API for creating and managing the user interface elements (e.g., windows, buttons, menus) and (b) a window manager to allow users to operate and manage the applications through its own user interfaces.

Figure 5.1 illustrates the UI layer as part of the system software in many computing platforms. The user interacts with the window/GUI-based *applications* using various input and output *devices*. At the same time, aside from the general applications, the user interacts with the computer and manages multiple application windows/tasks (e.g., resizing, focusing, cutting and pasting, etc.) using the

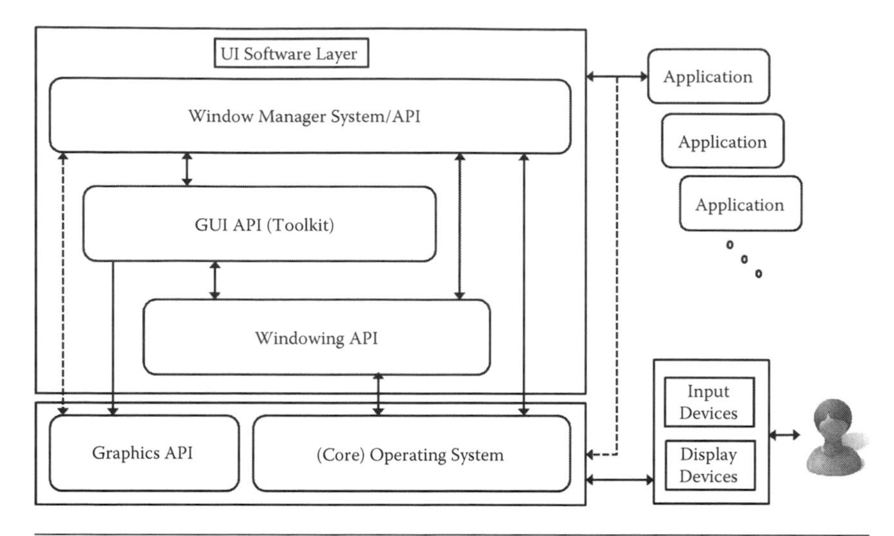

Figure 5.1 User interface software layer for a window-based multitasking UI.

(background running) *window manager*. The window manager is regarded as both an application and API. User applications are developed using the APIs that represent abstracted I/O-related functionalities of the UI layer, such as those for window managing (resizing, iconifying, dragging, copy and paste, etc.), GUI elements and widgets (windows, menus, buttons, etc.), and basic windowing (creating/destroying window, deactivating window, etc.). These APIs are abstracted from the even lower-level APIs for 2-D/3-D graphics and the operating system. Note that the architecture described here can be equally applied to nonwindow-based systems, such as those for layer-based systems* (e.g., mobile phones). Through such an architecture and abstraction, it becomes much easier to develop and implement interactive applications and their user interfaces.

5.2 Input and Output at the Low Level

At the lowest level, inputs and outputs are handled by the *interrupt* mechanism of the system software (operating system). An interrupt is a signal to the processor indicating that an event (usually an I/O event)

* The term *layer* refers to a full-screen interaction space, often used in small-display computing platforms. Multiple layers (representing multiple interactive processes) are possible. On smartphones, the layers are visible one at a time and can be switched by the use of swiping gestures on the touch screen.

has occurred and must be handled. An interrupt signal is interpreted so that the address of its handler procedure can be looked up and executed while suspending the ongoing process for a moment. After the handler procedure is finished, the suspended process resumes again. An arrival of an interrupt is checked at a very fast rate as part of the processor execution cycle. In practice, this means that the processor is always listening to the incoming events, ready to serve them as needed. The interrupt mechanism is often contrasted to *polling* (also known as *busy waiting*). In polling, the processor (rather than the I/O device) initiates input or output. In order to carry out I/O tasks, the processor enters a loop, continually checking the I/O device status to see whether it is ready, and incrementally accomplishes the I/O task. This form of an I/O is deficient in supporting asynchronous user-driven (anytime) I/O and wastes CPU time by blocking other non-I/O processes to move on.

At a higher level, the I/O operation is often described in terms of *events* and *event handlers*, which is in fact an abstraction of the lower-level interrupt mechanism. This is generally called the *event-driven architecture* in which programs are developed in terms of events (such as mouse clicks, gestures, keyboard input, etc.) and their corresponding handlers. Such information can be captured in the form of a table and used for efficient execution. Figure 5.2 shows the rather complicated interrupt mechanism abstracted into the form of a simple event-handler table.

5.3 Processing the Input and Generating Output

5.3.1 Events, UI Objects, and Event Handlers

The diagram in Figure 5.1 is the nominal software architecture for developing interactive computing systems. How does it work exactly? In other words, how is the user/device input processed, and how does the application (with the help of the UI layer) generate output? Central to its overall interworking are *events*, *UI objects*, and *event handlers*.

The most basic UI object in today's visually oriented UI system would be the *window* (or layer). A window is a rectangular portion of the screen associated with a given application that is used as a space and channel for interacting with the application. Other UI objects

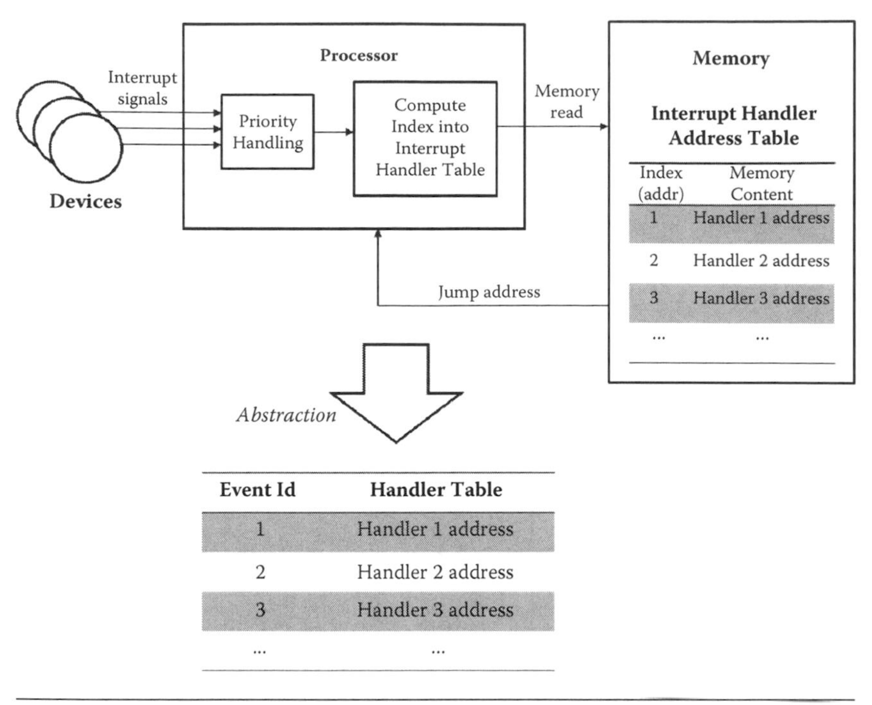

Figure 5.2 Complex interrupt mechanism abstracted as an event-handler table.

include buttons, menus, icons, forms, dialog boxes, text boxes, and so forth. These are often referred to as *GUI objects* or *widgets*. Most typically, GUI-based interactive applications would have a top window that includes all other UI objects or widgets that are logically and/or spatially subordinate to it (Figure 5.3). With the current operating systems mostly supporting concurrency, separate windows/widgets for concurrent applications can coexist, overlapping with one another so that they can be switched to the current focus. That is, when there are multiple windows (and one mouse/keyboard), the user carries out an action to designate the active or current window *in focus* to which the input events will be channeled. Two major methods for focusing are (a) click-to-type and (b) move-to-type. In the former, the user has to explicitly click on the window before making input into it (regardless of the mouse position, the last object that was clicked on will be the one in focus), and in the latter, the window over which the mouse cursor hovers becomes the focus. The move-to-type method is generally regarded as less convenient because of the likelihood of unintended focus change due to accidental mouse movements.

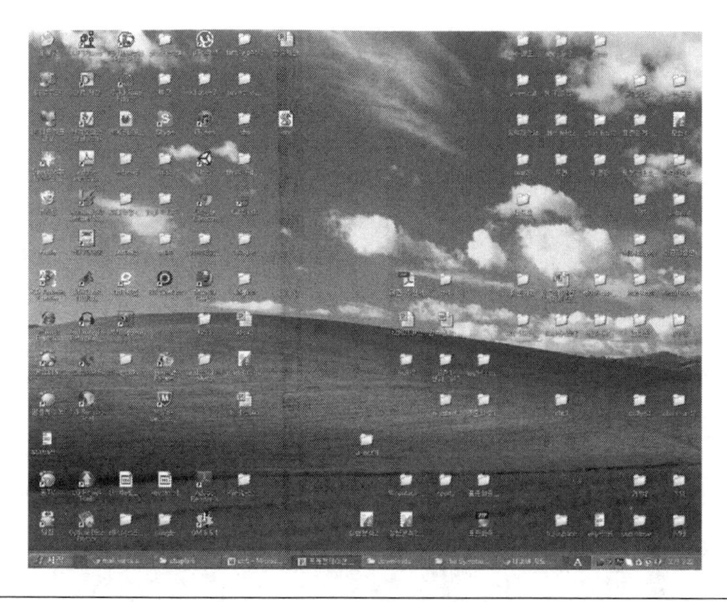

Figure 5.3 Root (background) window activated by default upon system start and also serving as the top window for the window manager process.

While not all UI layers are modeled and implemented in an object-oriented fashion, many recent ones are. Thus we can think of generic or abstract object classes for a window and other UI objects and widgets as being organized hierarchically (Figure 5.4). Moreover, we can designate the background screen space as the default root system window (which becomes automatically activated upon system start) onto which children application windows and GUI elements (e.g., icons, menus) are placed. The background also naturally becomes the top window for the window manager process.

Whether it is the root (background) window, application (top) window, or GUI widget, as an interaction channel or object, it will receive input from a user through input devices such as the keyboard, mouse, etc. The physical input from the user/devices is converted into an event (e.g., by the device drivers and operating system), which is simply data containing information about the user's intent or action. Aside from the event value itself (e.g., which key was pressed), an event usually contains additional information such as its type, a time stamp, the window to which it was directed, and screen coordinates (e.g., in the case of an event activated by a mouse or a stylus). These events are put into a queue by the operating system (or the windowing system) and dispatched (or dequeued), e.g., according to the current focus (to be

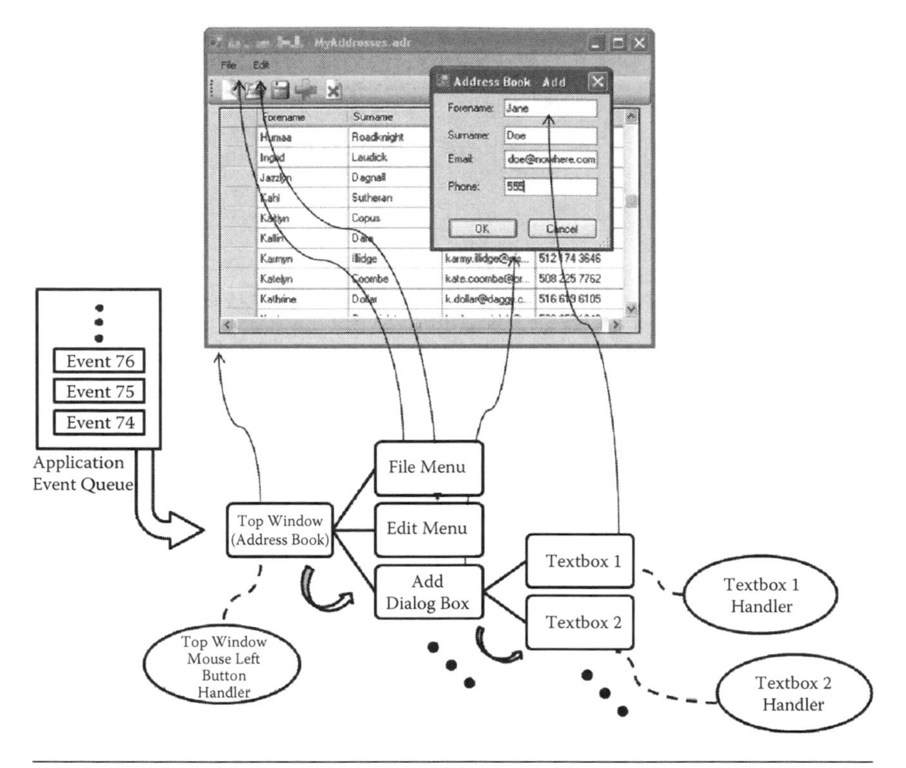

Figure 5.4 Event being dispatched to the right UI object handler for a given application (organized as a set of UI objects and associated event handlers in a hierarchical manner) from the application-event queue.

directed to the target program or process), to invoke its corresponding handler.

Note that an event does not necessarily correspond exactly just to an individual physical input. The stream of raw inputs may be filtered and processed to form/find meaningful input sequences from the current context. For instance, a sequence of raw inputs may form a meaningful event such as a double-click command, keyboard commands with modifiers (e.g., CTRL-ALT-DEL), and mouse enter/exit commands (e.g., detection of mouse cursor leaving a particular window).

Figure 5.5 shows the two-tier event-queuing system in greater detail. There is the system-level event-queuing system that dispatches the events at the top application level. Each application or process also typically manages its own event queue, dispatching them to its own UI objects. The proper event is captured by the UI object as it traverses down the application's hierarchical UI structure, e.g., from top to bottom. Figures 5.4 and 5.5 illustrate this process. Then the *event*

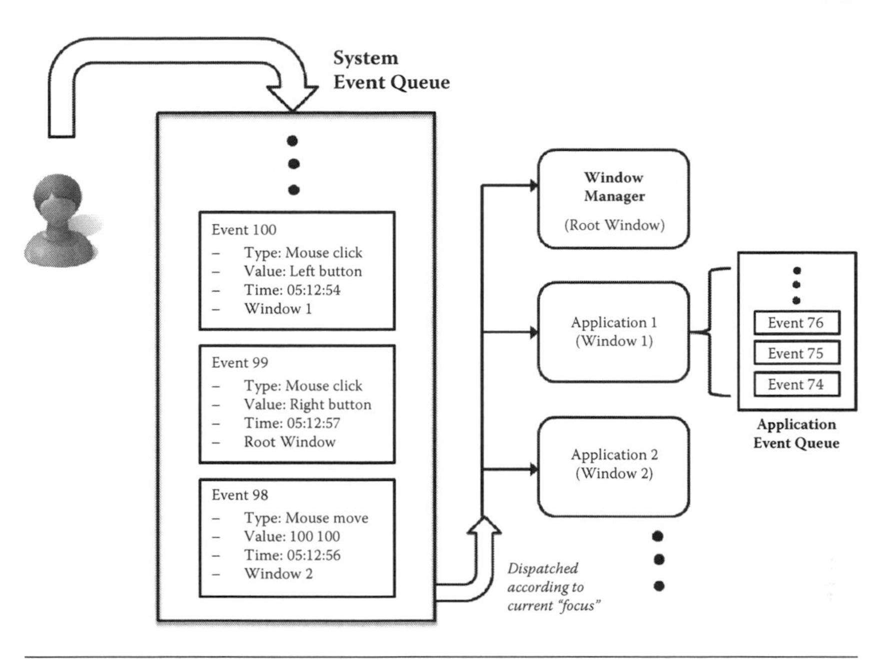

Figure 5.5 Event queuing at the top application level.

handler (also sometimes called the *callback function*) associated with the UI object is activated in response to the event that is captured.

The events do not necessarily have to be generated externally by the interaction devices; indeed, sometimes they are generated internally for special purposes (these are sometimes called the *pseudo-events*). For instance, when a window is resized, in addition to the resizing event itself, the internal content of the window must be redrawn, and the same goes for the other windows occluded or newly revealed by the resized window. Special pseudo-events are enqueued and conveyed to the respective applications/windows. In the case of resizing/hiding/activating and redrawing of windows, it is the individual application's responsibility, rather than the window manager's, to update its display contents, because only the respective applications have the knowledge of how to update their content. Thus a special redraw pseudo-event is sent to the application with information about which region is to be updated (Figure 5.6). The window content might need to be redrawn not because of the window management commands such as resizing and window activation, but due to the needs of the application itself, which can generate special pseudo-events for redrawing parts of its own window. More generally, UI objects can

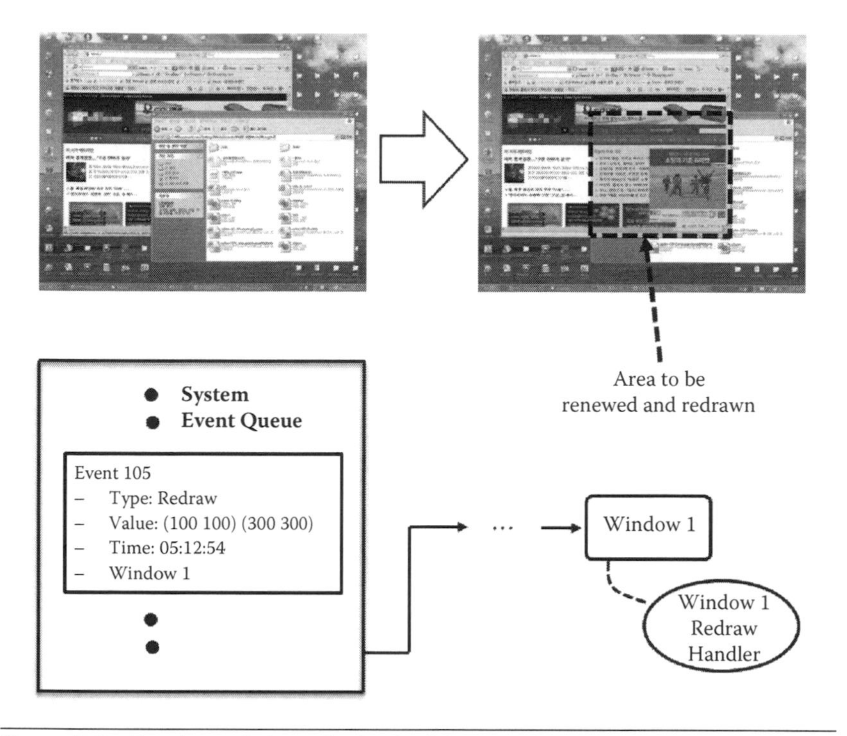

Figure 5.6 Exposing a window and redrawing it by enqueuing a special redraw event with the update area information. The event is matched to a proper redraw handler for the given application.

generate pseudo-events for creating chain effects. For example, when a scroll bar is moved, both the window content and the scroll bar position have to be updated [1].

5.3.2 Event-Driven Program Structure

Based on what we have discussed so far, the event-driven program structure generally takes the form of the structure depicted in Figure 5.7. The first initialization part of the program creates the necessary UI objects for the application and declares the event-handler functions and procedures for the created UI objects. Then the program enters a loop that automatically keeps removing an event from the application event queue and invoking the corresponding handler (i.e., dispatching the event). The development environment often hides this part of the program so that the developer does not have to worry about such system-level implementations. However, depend-

```
1. initialize application
 define UI objects
 define event handlers
 ...
 ...

2. run "system_loop"

system_loop ( ) {
 while (event != QUIT) {
  get next event from application queue
  find the target_object by traversing down the UI object tree
  invoke corresponding event handler for this target_object
  refresh screen and send commands to aural/haptic device
  }
} /* system provided event processing loop */
```

```
1. initialize application
 define UI objects
 define event handlers
 ...
 ...

2. while (event != QUIT) {
 get next event from application queue
 find the target_object by traversing down the UI object tree
 switch (event) {
   case mouse_right_button:
        /* compute for response behavior to this event for this object */
        /* compute for new or changes in visual/aural/haptic/tactile output */
        target_object.mouse_right_button_handler ( ... )

   case mouse_left_button:
        target_object.mouse_left_button_handler ( ... )

   ...
  } /* event dispatching */
 refresh screen and send commands to aural/haptic device
} /* user implemented event processing loop */
```

Figure 5.7 Event-driven program structure: UI object creation and event-handler setup followed by the event-processing loop either provided by the underlying programming environment system/operating system (above) or by explicit user programming (below).

ing on the development toolkits (see Chapter 6), the user may have to explicitly program this part as well.

5.3.3 Output

Interactive behavior that is purely computational will simply be carried out by executing the event-handler procedure. However, response to an event or application behavior is often manifested by explicit visual, aural, and haptic/tactile output as well. In many cases, the event handlers only compute for the response behavior and for needed changes in data or new output in a chosen modality (e.g., visual, aural, haptic, tactile, etc.). A separate step for refreshing the display based on the changed screen data is called as the last part of the event-processing loop. Analogous processes will be called for sending commands to output devices of other modalities as well (see the last line in Figure 5.7). Sometimes, with multimodal output, the outputs in different modalities need to be synchronized (e.g., output visual and aural feedback at the same, or nearly the same, time). However, not many interactive programming frameworks or toolkits offer provisions for such a situation.

While internal computation takes relatively little time (in most cases), processing and sending the new/changed data to the display devices can take a significant amount of time. For instance, a heavy use of 3-D graphic objects can be computationally expensive (e.g., on a mobile device without a graphics subsystem), and this can become a bottleneck in the event-processing loop, thereby reducing interactivity. Thus sometimes rendering and sensing parts can be separated into independent threads and processed at different rates to ensure real-time interactivity.

5.4 Summary

In this chapter, we looked at the inner workings of the general underlying software structure (UI layer or UI execution framework) on which interactive programs operate. Most UI frameworks operate in similar ways according to an event-driven structure. The hardware input devices generate events that are conveyed to the software interfaces (i.e., UI objects), and they are processed to produce output by the event-handling codes. The UI layer sitting above the operating system (OS) provides the computational framework for such an event-driven processing model and makes useful abstractions of the lower OS details for easier and intuitive interactive software and interface development. The next chapters introduce toolkits and development frameworks that make the interface development even more convenient and faster.

Reference

1. Olsen, Dan. 1998. *Developing user interfaces: Interactive technologies*. San Francisco, CA: Morgan Kaufman.

6
UI Development Toolkit

Now that we have the basic understanding of how the user-interactive (UI) software operates with *events* (most of them predefined and sometimes newly defined by the user), *UI objects*, and *event handlers*, the next question will be: "What are the *programming constructs* that allow us to specify these for actual implementation of a working interactive program?" As was mentioned in the early part of Chapter 4 and as illustrated in Figure 5.1, interactive programs and their interfaces are often developed using UI development toolkits. To be more precise, interfaces are developed using the UI toolkits and the core application logics using conventional programming languages. Obviously, the UI toolkits are closely related to the UI execution framework (Chapter 5) upon which the resulting interactive program would be running.

In the larger scheme of things, we can also think of a *UI development framework* (Chapter 7) as a methodology for the interactive program development as a whole. One such example might be a modular approach, where the core computational and interface parts are developed separately and combined in a flexible manner. This allows the concept of plugging in different interfaces for the same model computation and thus easier maintenance of the overall program. We first take a look at the UI toolkit.

6.1 User Interface Toolkit

The UI toolkit is a library of precomposed UI objects (which would include event handlers) and a predefined set of events that are defined and composed from the lower-level UI software layer or the UI execution framework. The UI toolkit abstracts the system details of handling events and, as such, programming for interactive software becomes easier and more convenient. The UI object often takes the form of a manipulable graphical object, often called a *widget* (i.e., *window gadget*). A typical widget set includes menus, buttons, text boxes, images, etc. We have already examined typical widgets and UI objects in Chapter 4, which showed that widgets may be singular or composite (made up of several UI objects). The use of a toolkit also promotes the creation of an interface with a consistent look, feel, and mechanism. Here, we take a closer look at the toolkits through three examples. In particular, we examine how events are defined, how UI objects are created, how event handlers are specified, and how the interface (developed this way) is combined with the core functional part of the application.

6.2 Java AWT UI Toolkit

Java, as an object-oriented language, offers a library of object classes called the AWT (Abstract Window Toolkit), which are classes useful for creating two-dimensional (2-D) UI and graphical objects [1]. The *component* is the most bare and abstract UI class from which other variant UI objects derive. Descendants (subclasses) from the component class include window, button, canvas, label, etc. The *window* class has further subclasses, such as *frame* and *dialog*. Each class has basic methods. For example, a window has methods for resizing, adding subelements, setting its layout, moving to a new location, showing or hiding, etc. Figure 6.1 shows the overall UI object hierarchy and an example of the codes for creating a frame (a window with a menu bar) and setting some of its properties by the calling of such methods.

The Java AWT is not just a library of object classes for programming, but also a part of the UI execution framework for Java that handles a (large) subset of interaction events (called the *AWTEvents*). In general, the interaction events are sent to Java programs, where they are captured, abstracted, and stored as *EventObjects*. The AWTEvents are descendants of the *EventObjects* that cover most of the useful UI events (such as mouse clicks, keyboard input, etc.). The AWT framework will map the AWTEvent to the corresponding AWT UI object. There are two ways for the UI object to handle the events.

The first is by overriding the (predefined) callback methods of the interactive applet object for the events. Table 6.1 shows the AWTEvent types and the corresponding callback methods that can be overridden for customized event handling. Figure 6.2 shows a code example for

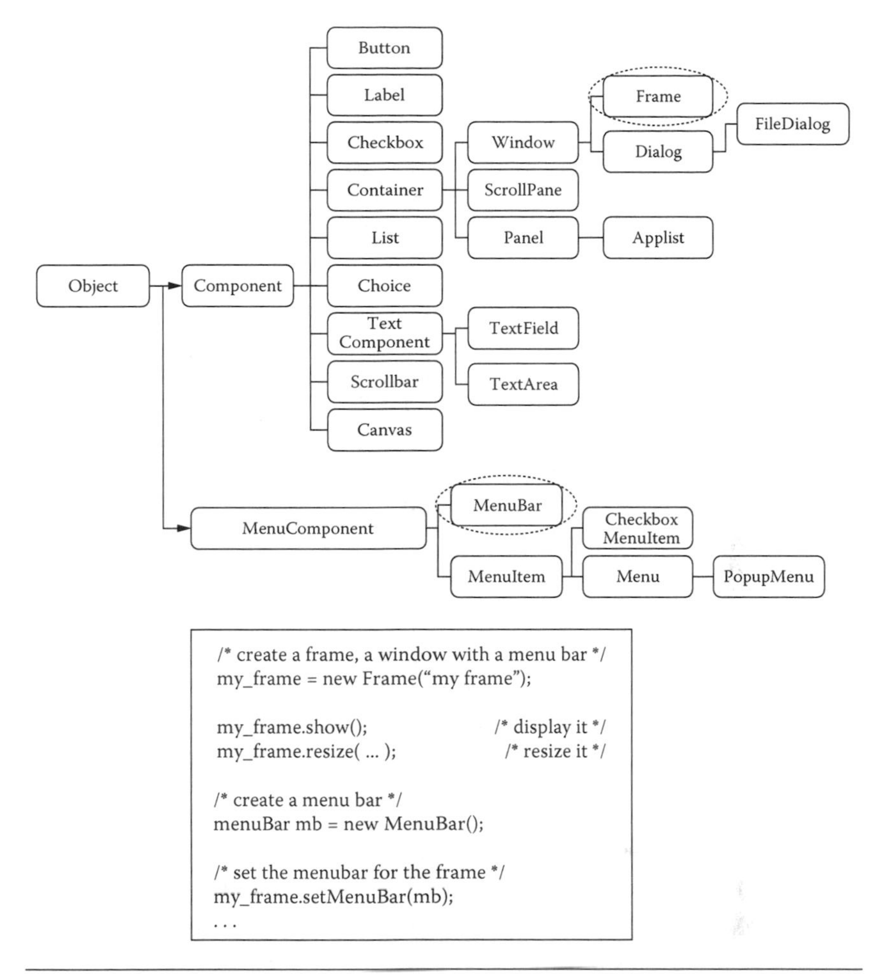

Figure 6.1 The class hierarchy of UI objects (top) and creating a window and setting some of its properties through calling its class-specific methods (bottom).

Table 6.1 AWTEvent Types and Corresponding Overridable Callback Functions

AWTEVENT TYPE	CALLBACK FUNCTION
mouseDown	mouseDown (Event evt, int x, int y)
mouseUp	mouseUp (Event evt, int x, int y)
mouseEnter	mouseEnter (Event evt, int x, int y)
mouseExit	mouseExit (Event evt, int x, int y)
mouseDrag	mouseDrag (Event evt, int x, int y)
gotFocus	gotFocus (Event evt, Object x)
lostFocus	lostFocus (Event evt, Object x)
keyDown	keyDown (Event evt, int key)
keyUp	keyUp (Event evt, int key)
action	Action (Event evt, Object x)

```
import java.awt.*;
import java.applet.*;

public class My_Applet extends Applet
{
        ...

        // Overridden methods for event handling
        public boolean mouseEnter( Event evt, int x, int y)
        {
                draw_object (x, y); // draw object at mouse clock position

                // Signal we have handled the event
                return true;
        }

        public boolean mouseMove( Event evt, int x, int y)
        {
                repaint()               // repaint whole screen
                draw_cursor (x, y);     // draw new cursor

                // Signal we have handled the event
                return true;

        }

        ...
}
```

Figure 6.2 An interactive applet with callbacks for mouse events. When the mouse click is entered, the applet draws an object at the click position. When the mouse moves, the whole applet is repainted and a new cursor is drawn at the newly moved position.

reacting to various mouse events and redrawing the given interactive applet. Note that each event handler must return a Boolean value (true or false), indicating whether the event should be made available to other event handlers. If, however, the event cannot be processed for some reason, the event handler may return *false* to signal that other components should process it.

As a second mechanism for implementing reactive behaviors to various events, the individual AWT UI object is to be registered with an event listener that waits for and responds to the corresponding event. The event handlers in Java AWT are known as the *listeners*.* As its name suggests, it is a background process that listens for the associated events for a given UI object and responds to them. It is an abstraction of the event-processing loop we have previously discussed in this section,

* To be precise, a *listener* is differentiated from a simple callback function because it is a process that waits for and reacts to the associated event, while a callback function is just the procedure that reacts to the event.

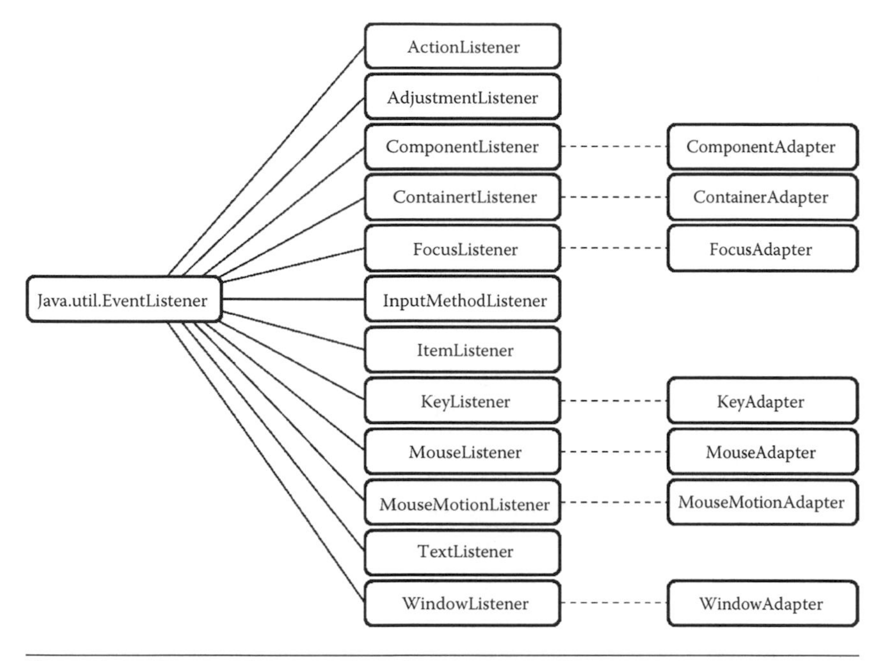

Figure 6.3 The event-listener interface hierarchy in Java AWT.

assigned to a UI component for each of various input events. Thus the listeners must be registered for various events that can be taken up by the given UI object. As a single UI object may be composed of several basic components and potentially receive many different types of input events, listeners for each of them will have to be coded and registered. Such a UI object is modeled as a collection of listeners through the Java implementation-interface construct (Figure 6.3).

Let's go back and take a look at the event-component hierarchy (Figure 6.4). Table 6.2 shows more-detailed descriptions of some of the various events. All the events derive from an abstract *EventObject* and offer basic methods for retrieving the object associated with the event and accessing the event type and identification (ID). Descendant

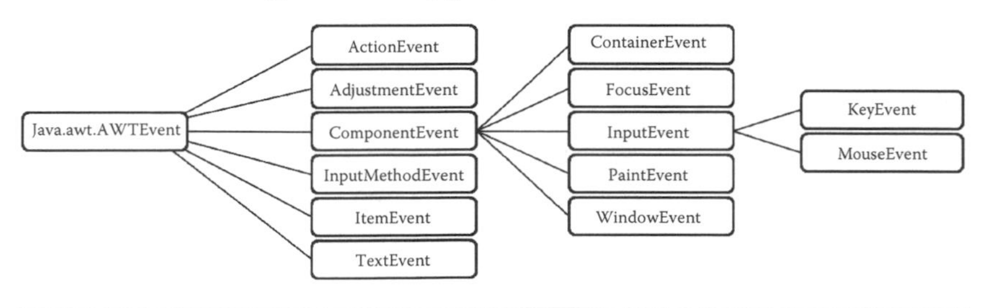

Figure 6.4 The event-component hierarchy in Java AWT.

Table 6.2 Java AWT Event Description and Examples

EVENT CLASS	DESCRIPTION/EXAMPLES
ActionEvent	Button press, double-click on an item, selection of a menu item
AdjustmentEvent[a]	Scroll bar movement
ComponentEvent	Hiding/revealing a component, component movement, and resizing
FocusEvent	Component gaining or losing a focus
KeyEvent	Keyboard input
ItemEvent	Checkbox selection/deselection, menu item selection
MouseEvent	Mouse button, mouse moving, mouse dragging, mouse focus
TextEvent	Text entry
WindowEvent	Window opening and closing, window activation/deactivation, window iconification/deiconification

[a] See http://docs.oracle.com/javase/6/docs/api/java/awt/event/AWTEventListenerProxy.html.

event classes possess additional specific attributes and associated methods for accessing the values. For instance, the *KeyEvent* has a method called *getkeyChar* that returns the value of the keyboard input; the *MouseEvent* has methods called *getPoint()* and *getClickCount()* that return the screen-position data at which the mouse event occurred and the number of clicks. For more detailed information, the reader is referred to the reference manual for Java AWT [1].

Just like the event-component hierarchy, the event-listener interface is also structured correspondingly, as seen in Figure 6.3. A UI object, possibly composed of several basic UI components, is designated to react to different events by associating the corresponding listeners with the UI object by the interface-implementation construct. That is, the UI object is declared to implement the various necessary listeners, and in the object-initialization phase, the specific components are created and listeners are registered. The class definition will thus include the implementation of the methods for the registered listeners. The named methods for the various listeners are illustrated in Table 6.3.

An example of a UI object as an event-listener implementation with some of its associated methods is shown in Figure 6.5. That is, *my_UI_object* is declared as an extension of an applet and at the same time as an implementation of two listeners (reacting to three types of events): the *ActionEvent*, and *MouseEvent*. It also has two button components created in the *init()*. Two listeners are also registered to each of the buttons (b1 and b2) in the same *init()* method. A description of the method implementations follows.

Table 6.3 Events, Corresponding Listener Interfaces, and Derived Methods in Java AWT

EVENT CLASS	CORRESPONDING LISTENER DESCRIPTION	SAMPLES OF DERIVED METHODS
ActionEvent	ActionListener	actionPerformed
AdjustmentEvent	AdjustmentListener	adjustmentValueChanged
ComponentEvent	ComponentListener	componentHidden componentMoved componentResized
FocusEvent	FocusListener	focusGained focusLost
KeyEvent	KeyListener	keyPressed keyReleased keyTyped
ItemEvent	ItemListener	itemStateChanged
MouseEvent	MouseListener	mouseClicked mouseEntered mouseExited mousePressed mouseReleased
MouseMotionEvent	MouseMotionListener	mouseDragged mouseMoved
WindowEvent	WindowListener	windowOpened windowClosed windowActivated windowDeactivated windowDeiconified windowIconified

6.3 Android™ UI Execution Framework and Toolkit

The user programming environment and execution model for Android (even though at the low level, the operating system is derived from Linux) is based on Java [2]. As such, the Android event-processing model and programming toolkit structure are mostly the same as those of Java (or more specifically Java AWT), except that the Android UI toolkit, in addition to the programmatic method, includes a declarative one for specifying the UI and defining its behaviors.

Events in Android can take a variety of different forms, but they are usually generated in response to bare and raw external actions, such as touch and button input. Multiple or composite higher-level events may also be internally recognized and generated, such as touch gestures (e.g., flick, swipe) or virtual-keyboard inputs. The Android framework maintains an event queue into which events are placed

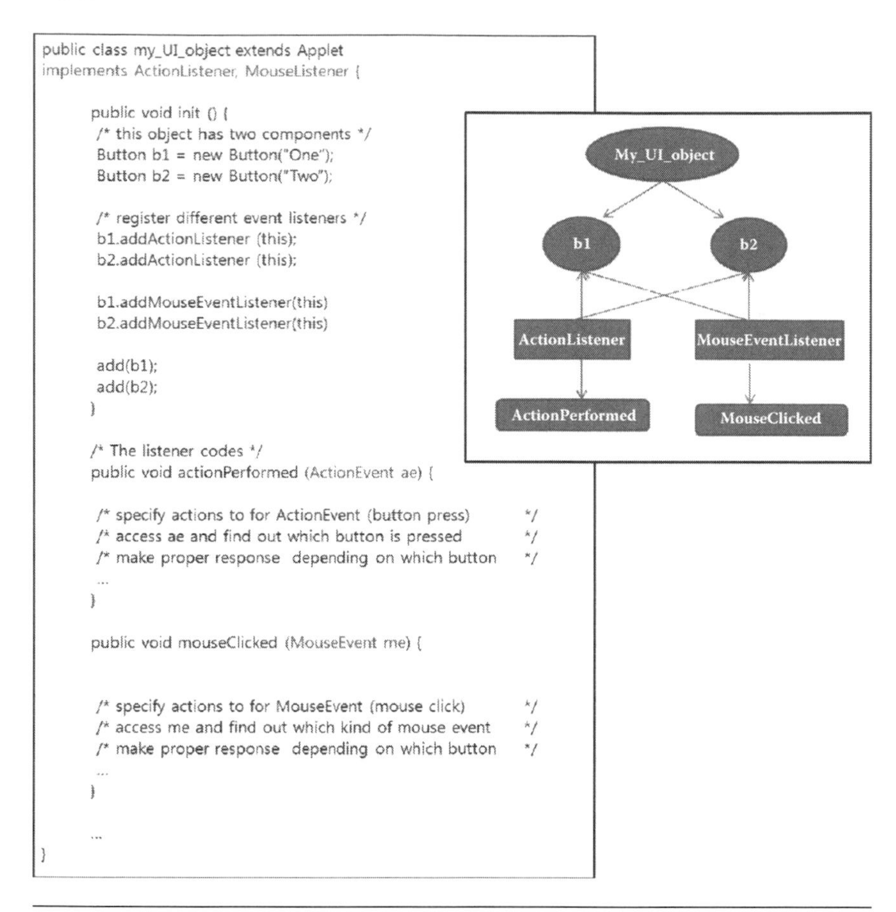

Figure 6.5 A UI object specification with event listeners using the Java AWT.

as they occur. Events are then removed from the queue on a first-in, first-out (FIFO) basis. In the case of an input event such as a touch on the screen, the event is passed to the *View* object (the UI object classes in Android derive from what is called the *View* object), either by the location on the screen where the touch took place or by the current focus. In addition to the event notification, the view is also passed a range of information (depending on the event type) about the nature of the event, such as the coordinates of the point of contact between the user's fingertip and the screen.

Similar to the case of Java AWT, there are two major ways to define the reactive behavior to these events. The first is to override the default callback methods (Figure 6.6), similar to those in Table 6.3 for Java AWT, of the *View* interactive class object for various typical input events.

```
public class My_Activity extends Activity {
  ...

  protected class My_View extends View {
    ...

   public boolean onTouchEvent (MotionEvent event) {
     super.onTouchEvent (event);
     if (event.getAction() == MotionEvent.ACTION_DOWN) {
       ...
     }
   } /* overriden onTouchEvent */

   public onKeyDown(int keyCode, KeyEvent event) {
     super.onKeyDown (keyCode, event);
     if (keyCode == ...) {
       ...
     }
   } /* overriden onKeyDown */

  ...

  } /* extended My_View */
}
```

Figure 6.6 Overriding the default View methods such as the onTouchEvent and onKeyDown for defining interactive behaviors for the UI object My_View to the touch event and (virtual) keyboard inputs in the Android UI toolkit.

The second method is to associate an event listener with the View object. The Android View class, from which all UI components are derived, contains a range of event-listener interfaces, each of which contains an abstract declaration for a callback method. In order to respond to an event of a particular type, a view must register the appropriate event listener and implement the corresponding callback. For example, if a button is to respond to a mouse-click event (equivalent to the user touching and releasing the button view as though clicking on a physical button), it must both register the *View.OnClickListener* event listener (via a call to the target view's *Set.OnClickListener()* method) and implement the corresponding *onClick()* callback method. In the event that a click event is detected on the screen at the location of the button view, the Android framework will call the *onClick()* method of that view when that event is removed from the event queue. It is, of course, within the implementation of the *onClick()* callback method that any tasks should be performed or other methods called in response to the button click.

Figures 6.7–6.9 show three different (but in effect equivalent) ways to register and implement an event listener:

```
public class My_Activity extends Activity {

  /* Called when the activity is first created */
  @Override
  public void onCreate (Bundle savedInstanceState) {
     ...
     /* create view object */
     View myView = new View(this);

     /* associate a listener */
     myView.setOnTouchListener (my_touchListener);
  }

 /* extend the listener class and implement the handler method */
  class MyTouchListenerClass implements View.OnTouchListener {

     public boolean onTouch (View v, MotionEvent event) {
       ...
       if(event.getAction() == MotionEvent.ACTION_DOWN) {
         return true;
       }
       return false;
     }
  }

  /* create an instance of MyTouchListenerClass */
  MyTouchListenerClass my_touchListener = new MyTouchListenerClass();
}
```

Figure 6.7 MyActivity creates a View object, called *myView*, and redefines MyTouchListenerClass by extending the View.OnTouchListener. Then an instance of MyTouchListenerClass—my_touchListener—is created and registered to the listener for myView.

```
public class My_Activity extends Activity {
   /** Called when the activity is first created. */
   @Override
   public void onCreate(Bundle savedInstanceState) {
      ...
      MyView myView = new MyView(this);
      myView.setOnTouchListener(myView);
      ...
   }

   protected class MyView extends View implements
View.OnTouchListener {
      ...

      public boolean onTouch (View v, MotionEvent event) {
         // TODO Auto-generated method stub
         return false;
      }
   }
   ...
}
```

Figure 6.8 In this example, My_Activity creates MyView class by implementing the View. OnTouchListener and instantiates a MyView object vw. The listener is registered by calling the Set. OnTouchListener on itself.

```
public class My_Activity extends Activity implements View.OnTouchListener {
   /** Called when the activity is first created. */
   @Override
   public void onCreate(Bundle savedInstanceState) {
      ...
      View myView = new View(this);
      myView.setOnTouchListener(this);
      ...
   }

   public boolean onTouch (View v, MotionEvent event) {
      if(event.getAction() == MotionEvent.ACTION_DOWN) {
         return true;
      }
      return false;
   }
   ...
}
```

Figure 6.9 In this example, My_Activity implements the View.OnTouchListener and thus includes and overrides the onTouch handler.

1. Implementing the event listener itself and associating it with the *View* object (which is often part of the Android *Activity* object, Figure 6.7)
2. Having the *View* object implement the event listener (Figure 6.8)
3. Having the topmost *Activity* object (which houses various view objects as its parts) implement the event listener (Figure 6.9).

As already mentioned, the Android UI framework also provides a declarative method for specifying the UI. That is, the form of the UI can be "declared" using a markup language (Figure 6.10). Through a development tool such as Eclipse [3], the UI can be built through direct graphical manipulation as well. Thus, in summary, there are three methods of UI development: (a) the usual programmatic way, (b) a declarative way, and (c) a graphical way. In Eclipse, the de facto development tool for Android applications, any methods can be used, and the three methods can even be used simultaneously. Figure 6.10 shows the declarative UI specification (main.xml file) for the No Sheets application designed in Chapter 4 through one of the subwindows in Eclipse.

The corresponding UI can be displayed in graphic form and be manipulated as shown in Figure 6.11. When a UI component is added or deleted or when an attribute value changes, such actions are reflected back to the respective representations, either graphic or declarative. The figures show that the UI screen is composed of an image and several buttons. The declarative specification names the

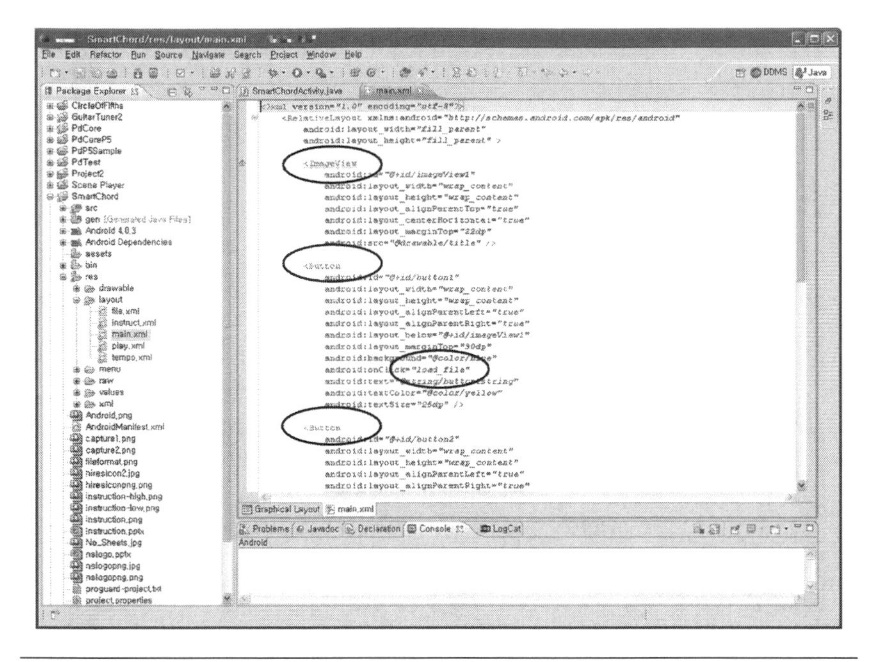

Figure 6.10 An example of a declarative specification of the UI for the No Sheets application (Chapter 4). The declarative specification is saved in an XML file with element constructs for UI components such as image, buttons, etc. Note that the file names the elements (for later referral in the program) and that one of the attributes of the UI object is the UI handler name (e.g., the encircled load_file).

event handler and other attribute values for the components in more exact terms. The handler code is implemented in the corresponding programmatic representation (Figure 6.12).

6.4 Example: iOS UIKit Framework and Toolkit

There are three major types of discrete events for iOS: multitouch, motion, and remote control (discrete events from external devices, such as remote-controlled headphones) [4]. The iOS generates low-level events when users touch "views" of an application. The application sends these discrete events as *UIEvent* objects, as defined by the *UIKit* framework, to the view (i.e., specific UI component) on which the touches occurred. The view analyzes the touch event and responds to them. Touch events can be combined to represent higher-level gestures such as flicks and swipes. The given application uses the *UIKit* classes for gesture recognition and responding to such recognized events. For continuous streams of sensor data such as those from

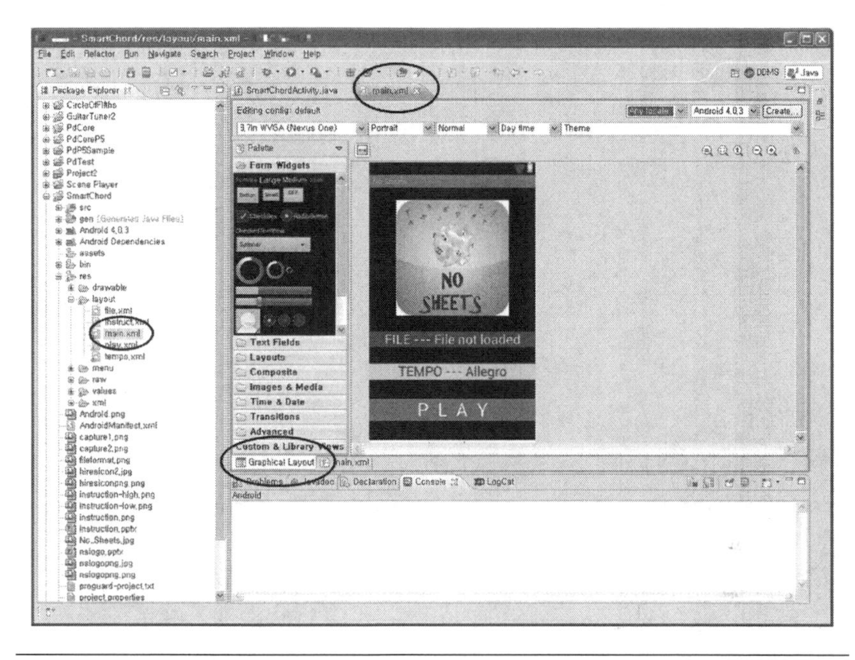

Figure 6.11 An example of a graphical specification of the UI for the No Sheets application (Chapter 4). The graphically displayed UI is consistent with the corresponding declarative representation (saved in the file main.xml).

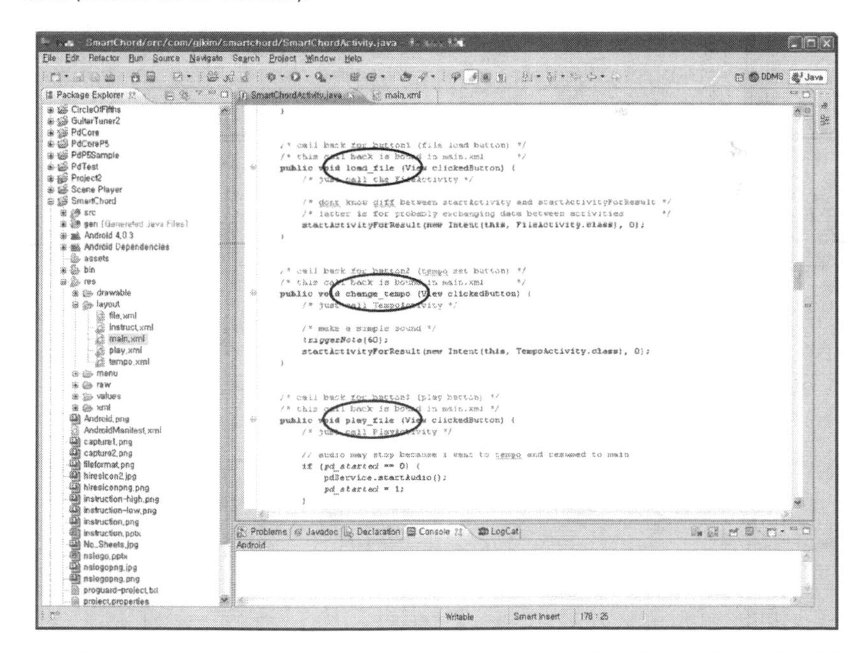

Figure 6.12 An example of a programmatic specification of the UI for the No Sheets application (Chapter 4). The figure shows the code implementation for the UI handlers (e.g., load_file) defined in the declarative specification.

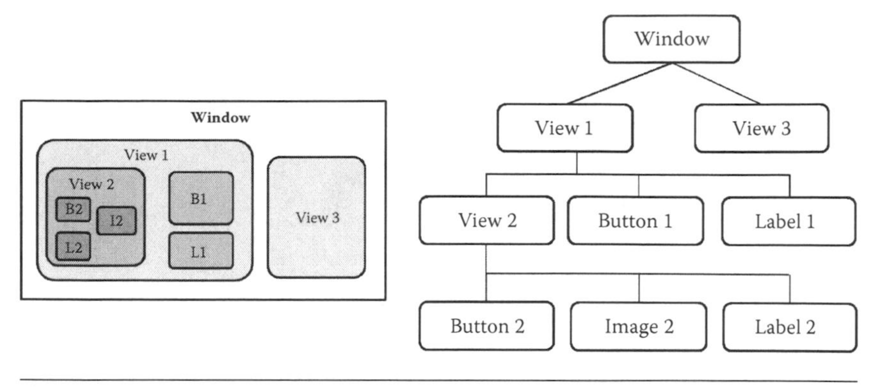

Figure 6.13 iOS UIKit A UI object hierarchy example.

accelerometers or gyroscopes, a separate *Core Motion* framework is used. Nevertheless, the mechanism is still similar in the sense that that sensor data, abstract event, or recognized event is conveyed from iOS to the application and then to the particular view according to the hierarchical structure of the application UI (Figure 6.13).

When users touch the screen of a device, iOS recognizes the set of touches and packages them in a *UIEvent* object and places it in the active application's event queue. If the system interprets the shaking of the device as a motion event, an object representing that event is also placed in the application's event queue. The singleton *UIApplication* object managing the application takes an event from the top of the queue and dispatches it for handling. Typically, it sends the event to the application's key window (the window currently in focus for user events), and the window object representing that window sends the event to an initial object for handling (Figure 6.14).

That object is different for touch events and motion events. For touch events, the window object uses hit testing and the "responder" chain to find the view to receive the touch event. In hit testing, a window calls *hitTest:withEvent:* on the topmost view of the view hierarchy; this method proceeds by recursively calling *pointInside:withEvent:* on each view in the view hierarchy that returns a yes, proceeding down the hierarchy until it finds the subview within whose bounds the touch took place. That view becomes the hit-test view. If the hit-test view cannot handle the event, the event travels up the responder chain until the system finds a view that can handle it. For Motion and Remote Control events, the window object sends each shaking-motion or remote-control event to the first responder for handling. Although the

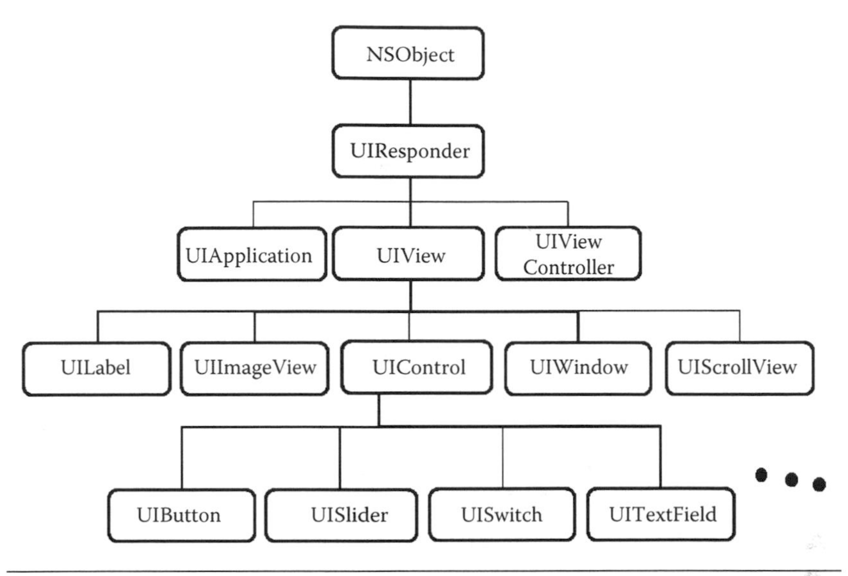

Figure 6.14 iOS UI event responder (handler) object class hierarchy. (From IOS Developers Library, iOS UIKit Framework Reference, 2013, https://developer.apple.com/library/ios/documentation/UIKit/Reference/UIKit_Framework/_index.html [4].)

hit-test view and the first responder are often the same view object, they do not have to be the same. The *UIApplication* object and each *UIWindow* object dispatch events in the *sendEvent:* method.

A responder object is an object that can respond to events and handle them. *UIResponder* is the base class for all responder objects, also known simply as responders. It defines the programmatic interface not only for event handling, but for common responder behavior. *UIApplication*, *UIView*, and all *UIKit* classes that descend from *UIView* (including *UIWindow*) inherit directly or indirectly from *UIResponder*, and thus their instances are responder objects. The first responder is the responder object in an application (usually a *UIView* object) that is designated to be the first recipient of events other than touch events. A *UIWindow* object sends the first responder these events in messages, giving it the first shot at handling them.

If the first responder or the hit-test view does not handle an event, *UIKit* may pass the event to the next responder in the responder chain to see if it can handle it. The responder chain is a linked series of responder objects along which an event is passed. It allows responder objects to transfer responsibility for handling an event to other, higher-level objects. An event proceeds up the responder chain as the application

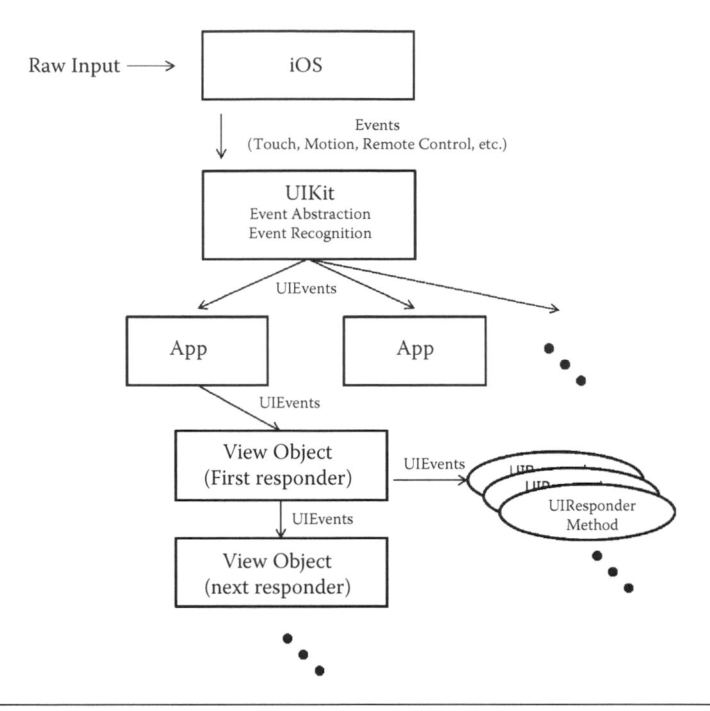

Figure 6.15 The event-processing flow and the event-driven object behavior structure. The user input is captured, abstracted, and recognized by the UIKit framework and queued into the proper application view objects (or responders), which implement the particular response behaviors using the UIResponder methods.

looks for an object capable of handling the event (Figure 6.15). Because the hit-test view is also a responder object, an application may also take advantage of the responder chain when handling touch events. The responder chain consists of a series of next responders (each returned by the nextResponder method) in the sequence. The response behavior itself is implemented by the responder objects (i.e., the UI components such as the window, button, slider, etc.).

6.5 Summary

In this chapter, we reviewed three examples of UI toolkits, namely, those for Java3D, Android, and iOS. There are certainly many other UI toolkits; however, most of them are similar in their structures and basic underlying mechanisms. As you have seen, some UI toolkits include visual prototyping tools and declarative specification syntax as well, which make it even more convenient for developers to implement user interfaces. In general, the use of UI toolkits promotes standardization, familiarity, ease of use, fast implementation, and consistency for a given platform.

References

1. Oracle. 2013. Abstract Window Toolkit. http://docs.oracle.com/javase/7/docs/api/java/awt/package-summary.html.
2. Google Developer. 2013. User interface. http://developer.android.com/guide/topics/ui/index.html.
3. Eclipse. 2013. http://www.eclipse.org/.
4. IOS Developer Library. 2013. iOS UIKit Framework Reference. https://developer.apple.com/library/ios/documentation/UIKit/Reference/UIKit_Framework/_index.html.

7 INTERACTIVE SYSTEM DEVELOPMENT FRAMEWORK

So far, we have only focused on the user interface (UI) objects and their behavior. Obviously, a complete application consists not only of UI objects, but those for the core functions of the application as well. How do we effectively develop the larger interactive programs with two such parts (i.e., UI and internal functional core)? For this, it is a good idea to follow an established development framework or methodology suited for highly interactive systems [1]. A *development framework* refers to a modular approach for interactive program development where the core computational and interface parts are developed in a modularized fashion and combined in a flexible manner. Such a development framework is often based on the UI toolkit, which provides the abstraction for the interface parts. For one, the framework allows the concept of plugging in different interfaces for the same model computation and easier maintenance of the overall program. In addition, such a practice also promotes the productivity as well as easier and less costly postmaintenance. MVC (model, view, and controller) is one such major framework.

7.1 Model, View, and Controller (MVC)

The MVC approach was first proposed as a computational architecture for interactive programs (rather than a methodology) by the designers of the programming language called SmallTalk, which is one of the first object-oriented and modular languages [2]. The modular nature of the MVC architecture naturally shaped the interactive program development style or methodology. With the MVC framework, the application is divided into three parts: (a) model, (b) view, and (c) controller, as illustrated in Figure 7.1.

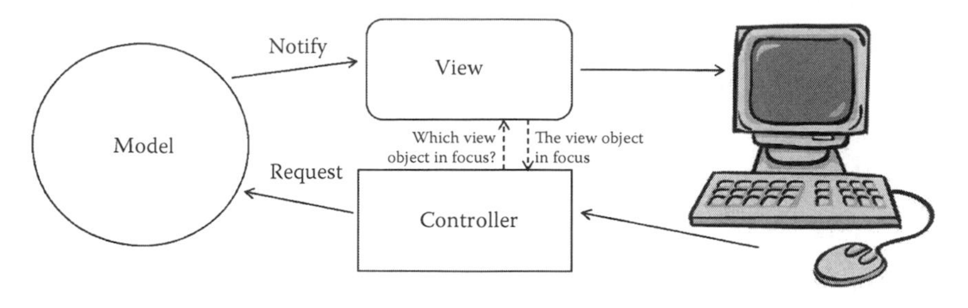

Figure 7.1 The MVC architecture for interactive applications.

7.1.1 Model

The *model* part of the application corresponds to the computation (e.g., realized as objects) that deals with the underlying problem or main information or data of the application. For all practical purposes, once in place, a model of the application tends to be stable and unchanging. For instance, in an interactive banking application, the model will be parts of the program that maintain the balance, compute the interest, make wire transfers, etc. The model has no knowledge of how the central information will be presented to the user (output/presentation) or how the transactions (input) are made.

7.1.2 View

The *view* part of the application corresponds to the implementation for output and presentation of data. In modern GUI-based interfaces, the implementation will typically consist of widgets. For instance, views might be windows and widgets that display the list of transactions and the balance of a given account in a banking application, or they might play a background audio clip depending on the score level for a game. As a whole, there may be multiple views for a single application (or model). For instance, there could be different view implementations for different display platforms or user groups (e.g., 17-in. monitor, 10-in. LCD, HD resolution display, display with vibrotactile output device, young users, elderly users). Note that the output display does not necessarily have to be visual.

Anytime the model is changed, the view of that model must be *notified* so that it can change the visual representation of the model on the output display. The region/portion of the screen/display that is no longer consistent with the model is said to be *damaged*. Oftentimes, it is too tedious to update just the damaged part of the display upon change of information in the model. The practical approach is to redraw the entire content of the smallest widget that encompasses the damaged region or redraw the entire window.

7.1.3 Controller

The *controller* part of the application corresponds to the implementation for manipulating the view (in order to ultimately manipulate the internal model). It takes external inputs from the user and then interprets and relays them to the model. The controller thus practically takes care of the input part of the interaction. It uses the underlying UI execution framework or operating system to achieve this purpose (while the view is mostly independent from the operating system or platform).

In Chapter 6, we studied the mechanism of the UI execution framework in terms of how it identifies and maps the raw user input to the object in focus. In order to find the object in focus (i.e., the visual object that is to be manipulated on behalf of the model), the controller must communicate with the view objects. In addition, the controller sometimes might also change the content of the display without changing the model. For instance, if the user wanted to simply change the color of a button (e.g., for UI customization purpose), the controller can directly communicate with the view to achieve this effect.

Once the object in focus is identified, the corresponding event handler would be invoked. The controller will only relay a query or message for a certain change or manipulation to happen to the model rather than actually making the change itself.

7.1.4 View/Controller

In many application architectures, the view and controller may be merged into one module or object because they are so tightly related to each other. For instance, a UI button object will be defined by attribute parameters such as its size, label, and color as well as the event handler that invokes the methods on the model for change or manipulation.

The MVC architecture or development methodology makes it much easier, particularly for large-scale systems, to quickly explore and implement and modify various user interfaces (view/controller) for the same core functional model. This is based on a famous software engineering principle: the separation of concern.

7.2 Example of MVC Implementation 1: Simple Bank Application

We illustrate a very simple object-oriented implementation for an interactive banking application. In this simple application, the model maintains the balance for a user who can make deposits or withdrawals through a computer UI. Figure 7.2 shows the overall structure of the application according to the MVC architecture.

In Figure 7.2, as for the model part, a class called *Account* maintains the customer name, balance, and two views/controllers (one for displaying the balance and realizing the UI for making deposits, and the other for withdrawal). The model has two core methods for maintaining the correct balance when a deposit (*Account::Deposit*) or withdrawal (*Account::Withdrawal*) is made. These two methods use the *Notify_depositVC* and *Notify_withdrawalVC* to notify the corresponding view to update the balance in the display.

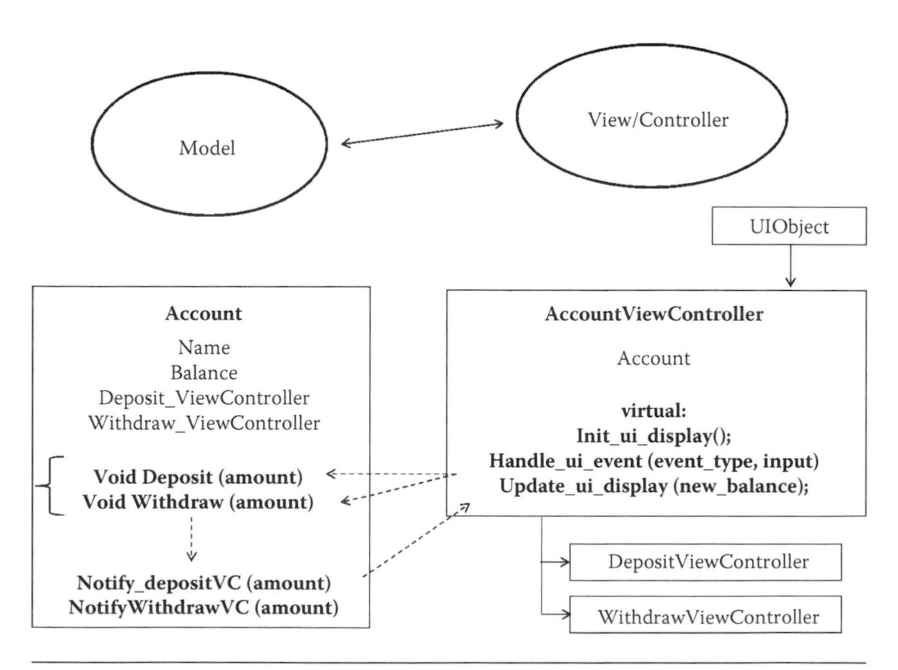

Figure 7.2 An overall MVC-based implementation (class diagram) for a simple interactive banking application.

In this particular example, the View and Controller parts are merged into one class, called the *AccountViewController*, which has a pointer to the corresponding model object (as the recipient of the notifications and model change queries). This class is a subclass of a more general *UIObject* that is capable of housing constituent widgets and reactive behavior to external input. It is also the superclass for subclasses, the *DepositViewController* and *WithdrawalViewController*, which implement the two views/controllers for the given model.

The subclasses, among others, implement three important virtual methods: *Init_ui_display*, *Update_ui_display*, and *Handle_ui_event* (Figure 7.2). Each of these is responsible for creating and initializing the display and UI objects within the view/controller, updating the display (invoked by the notification method from the model, as seen in Figure 7.3) and handling the user input. Figure 7.4 shows the *Handle_ui_event* method of the *DepositViewController* that interprets the user input (e.g., textual input of digits into integers) and invokes the model method, for example, to make a deposit by calling *my_model->Deposit(deposit_amount)*. Understand that this will eventually change the model, and the view/controller will be notified to change its display (e.g., to show the proper amount of balance after the deposit). Although not shown, the *WithdrawalViewController* would be coded in a similar manner.

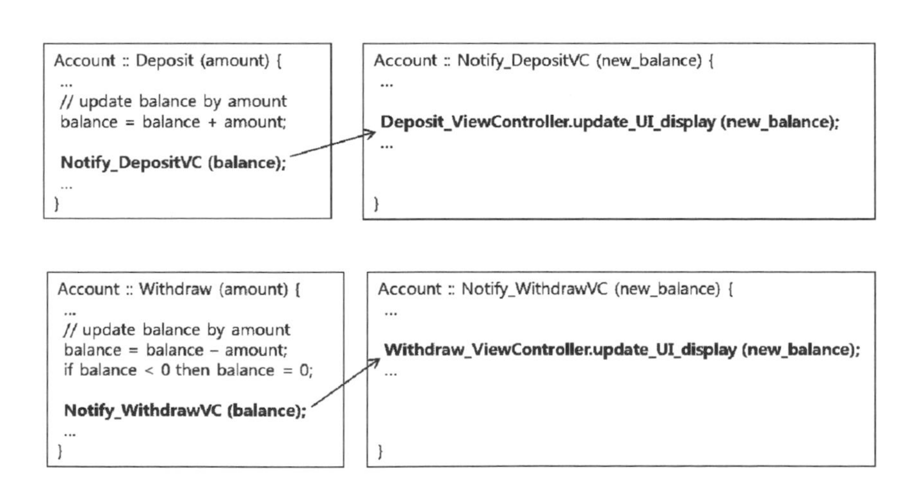

Figure 7.3 Three important virtual methods for the class Account (the Model).

```
DepositViewController :: AccountViewController {

 ...

 // Implement the subclass specific virtual methods

 // create and initialize UI display (the view) for deposit
 void Init_ui_display ();

 // redraw the UI display with new_balance
 void Update_ui_display (new_balance);

 // handle events from user (the controller)
 void Handle_ui_event (event_type, input);

}
```

```
DepositViewController :: Handle_ui_event (event_type, input) {

 ...

 // get the model
 my_model = Get_model ();

 // if the user input is text input
 if (event_type == TEXT_INPUT) {
        // convert the text input to integer
        deposit_amount = convert_to_int (input);
        // make a request to model to deposit
        my_model->Deposit (deposit_amount);
 }
 ...
}
```

Figure 7.4 The DepositViewController class and its method, Handle_ui_event.

7.3 Example of MVC Implementation 2: No Sheets

As a second example, we will illustrate parts of the implementation code for the No Sheets application introduced in Chapter 4, as shown in Figure 7.5. The core of the model is music information, a list-based data structure that contains the chord information for a piece of music that is read from a user-selected file. Aside from the music information itself, there may be other model variables such as the music file name, tempo value, etc. Thus, the model information is updated by and read from the view/controller objects.

The view/controller is composed of several *Activity* (screen interface) objects. The *SmartChordActivity* represents the main front-end interface, which allows the user to apply certain major actions such as selecting/loading the music file, selecting the tempo, playing the chosen music file, and other miscellaneous functions. It will access

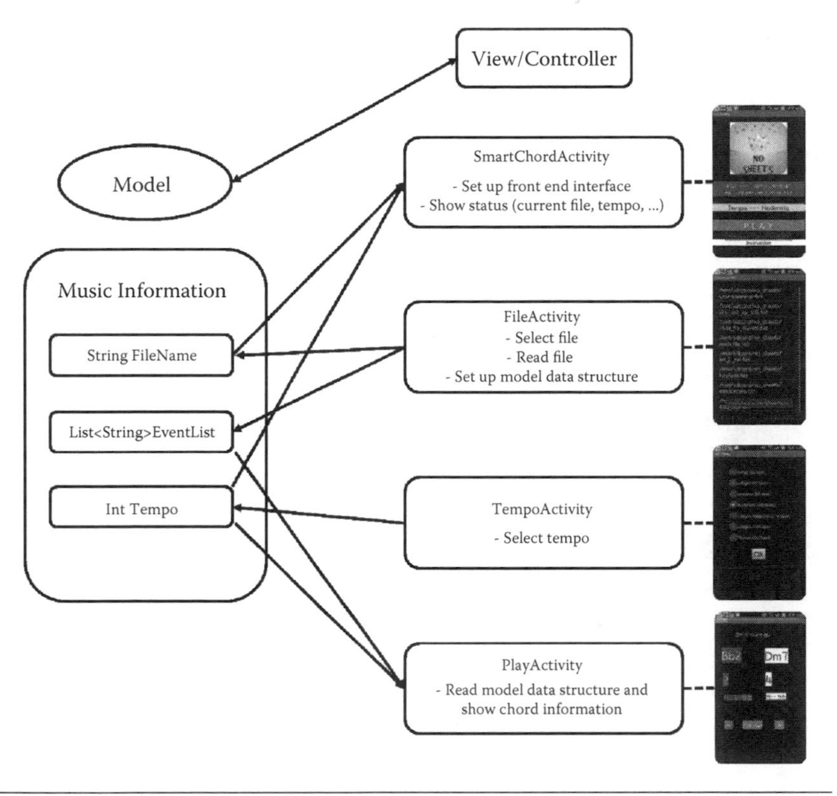

Figure 7.5 The MVC-based program structure for the No Sheets application introduced in Chapter 4.

information from the model—for instance, the name of the current file and current tempo—and show them in the interface. *FileActivity* represents the file-selection interface screen, which presents the user with a list of available music files. The user makes a selection, and the *FileActivity* will construct the internal chord-event list and update the model. Likewise, *TempoActivity* allows the user to select the tempo and update the model accordingly. Finally, the *PlayActivity* accesses the event-list data structure of the model and presents the musical information at a given tempo (no model updating is carried out).

7.4 Summary

In this chapter, we have studied one interactive application development methodology called the MVC, which is based on the principle of the separation between the UI and core computational functionalities of a given application. Such a separation of concerns allows for the two to be mixed and matched (for exploring different combinations of a proper set of functions and corresponding UIs) and lends itself to easier code maintenance. However, sometimes it is not very clear whether a given application can be cleanly separated into two parts, namely, the core function and UI. For example, suppose one is to implement several different "views" for different user groups for the same banking application, and yet another view for changing and selecting the views themselves. In this situation, it seems that the change-of-view functionality is one of the core functions and features of the application, yet in theory, the "view change" seems to belong to the View rather than the Model.

References

1. Olsen, Dan. 1998. *Developing user interfaces: Interactive technologies*. San Francisco, CA: Morgan Kaufman.
2. Krasner, Glenn E., and Stephen T. Pope. 1988. A cookbook for using the model-view controller user interface paradigm in SmallTalk-80. *Journal of Object-Oriented Programming* 1 (3): 26–49.

8
User Interface Evaluation

The last remaining part in the cycle of UI interactive software development is the evaluation stage. Even if the developers may have strived to adhere to various HCI principles, guidelines, and rules and have applied the latest toolkits and implementation methodologies, the resulting UI or software is most probably not problem-free. Frequently, careful considerations in interaction and interface design may not even have been carried out in the first place. Aside from the fact that there may be things that the developer failed to oversee or consider, the overall development process was to be a gradual refinement process to begin with, where the next refinement stages would be based on the evaluation results of the previous rounds. In this chapter, we will present several methods and examples of evaluation for user interfaces.

8.1 Evaluation Criteria

When evaluating the interaction model and interface, there are largely two criteria. One is the *usability* and the other is *user experience* (UX). Simply put, usability refers to the ease of use and learnability of the user interface (we come back to UX later in this section) [1]. Usability can be measured in two ways, *quantitatively* or *qualitatively*.

Quantitative assessment often involves task-performance measurements. That is, we assume that an interface is "easy to use and learn" (good usability) if the subject (or a reasonable pool of subjects) is able to show some (absolute) minimum user performance on typical application tasks. The assessment of a given new interface is better made in a comparative fashion against some nominal or conventional interface (in terms of relative performance edge). Popular choices of such performance measures are task completion time, task completion amount in a unit time (e.g., score), and task error rate. For example, suppose we would like to test a new motion-based interface for a smartphone game. We could have a pool of subjects play the game, using both the conventional touch-based interface and also the newly proposed motion-based one. We could compare the score and assess the comparative effectiveness of the new interface. The underlying assumption is that task performance is closely correlated to the usability (ease of use and learnability). However, such an assumption is quite arguable. In other words, task-performance measures, while quantitative, only reveal the aspect of efficiency (or merely the aspect of ease of use) and not necessarily the entire usability. The aspect of learnability should be and can be assessed in a more explicit way by measuring the time and effort (e.g., memory) for users to learn the interface. The problem is that it is difficult to gather a homogeneous pool of subjects with similar backgrounds (in order to make the evaluation fair). Measuring the learnability is generally likely to introduce much more biasing factors such as differences due to educational/experiential/cultural background, age, gender, etc. Finally, quantitative measurements in practice cannot be applied to all the possible tasks for a given application and interface. Usually, a very few representative tasks are chosen for evaluation. This sometimes makes the evaluation only partial.

To complement the shortcomings of the quantitative evaluation, *qualitative evaluations* often are conducted together with the quantitative analysis. In most cases, quantitative evaluations amount to conducting a usability survey, posing usability-related questions to a pool of subjects after having them experience the interface. A usability survey often includes questions involving the ease of use, ease of learning, fatigue, simple preference, and other questions specific to the given interface. NASA TLX (Task Load Index, Figure 8.1) and the IBM Usability Questionnaire (Figure 8.2) are examples of the often-used semi-standard questionnaires for this purpose [2, 3, 4].

User experience (UX) is the other important aspect of interface evaluation. There is no precise definition for UX. It is generally accepted that the notion of user experience is "total" in the sense that it is not just about the interface, but also something about the whole product/application and even extends to the product family (such as the Apple® products or MS Office). It is also deeply related to the user's emotions and perceptions that result from the use or anticipated use of the application (through the given interface) [4].

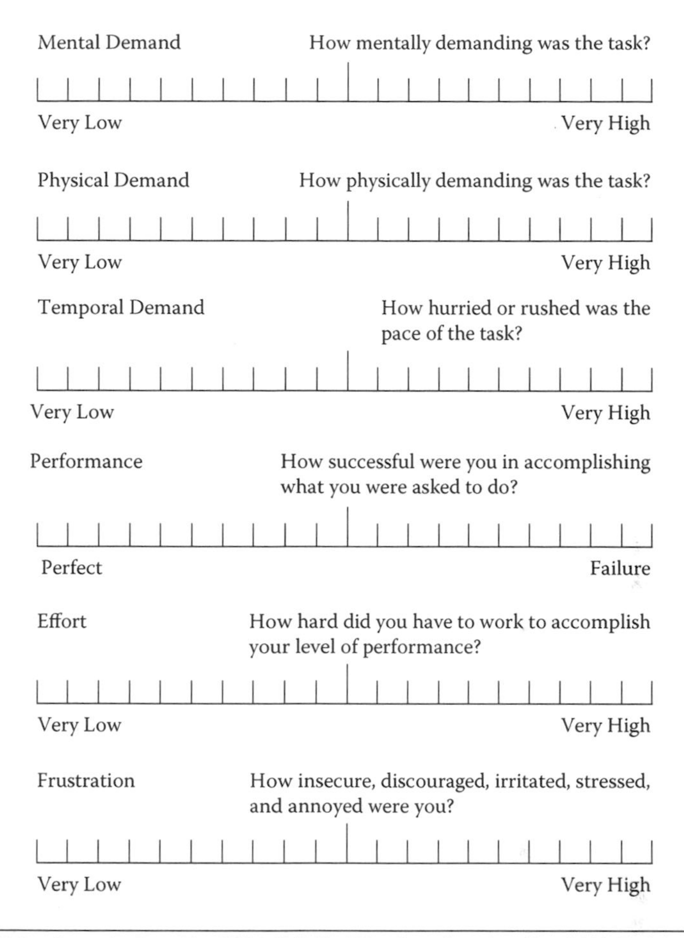

Figure 8.1 Excerpts from the NASA TLX Usability Questionnaire. The NASA Task Load Index method assesses the workload on a seven-point scale. Increments of high, medium, and low estimates for each point result in 21 gradations on the scale. (From Hart, S. G., Land, S., and Lowell, E., *Human Mental Workload,* 1(3), 139–83, 1988 [2]; NASA, NASA Task Load Index, 2013, http://human-systems.arc.nasa.gov/groups/tlx/downloads/TLXScale.pdf [3].)

Such an affective response is very much dependent on the context of use. Thus UX evaluation involves a more comprehensive assessment on the emotional response, under a variety of usage contexts and across a family of products/applications/interfaces (see Figure 8.3). A distinction can be made between usability methods, which have the objective of improving human performance, and user experience methods, which have the objective of improving user satisfaction by achieving both the pragmatic and hedonic goals [5]. Note that the notion of UX includes usability, i.e., high UX usually translates to high usability and high emotional attachment.

1.	Overall, I am satisfied with how easy it is to use this system.							
Strongly Agree								**Strongly Disagree**
Comments:	1	2	3	4	5	6	7	

2.	It was simple to use this system.							
Strongly Agree	1	2	3	4	5	6	7	**Strongly Disagree**
Comments:								

3.	I could effectively complete the tasks and scenarios using the system.							
Strongly Agree	1	2	3	4	5	6	7	**Strongly Disagree**
Comments:								

4.	I was able to complete the tasks and scenarios quickly using this system.							
Strongly Agree	1	2	3	4	5	6	7	**Strongly Disagree**
Comments:								

Figure 8.2 Excerpts from the IBM Usability Questionnaire for computer systems. (From Lewis, J. R., *International Journal of Human–Computer Interaction*, 7(1), 57–78, 1995 [6].)

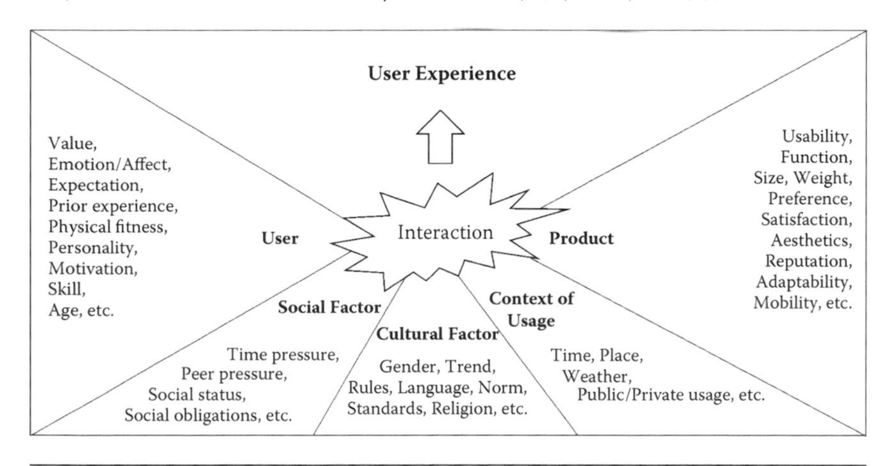

Figure 8.3 Various aspects to be considered in totality for assessing user experience (UX).

8.2 Evaluation Methods

Whether it is for the user experience or more narrow usability, or whether for the qualitative feelings or quantitative performance, there is a variety of evaluation methods. A given method may be general and applicable to many different situations and objectives, or it may be more specific and fitting for a particular criterion or usage situation. Overall, an evaluation method can be characterized by the following factors:

- Timing of analysis (e.g., throughout the application development stage: early, middle, late/after)
- Type and number of evaluators (e.g., several HCI experts vs. hundreds of domain users)
- Formality (e.g., controlled experiment or quick and informal assessment)
- Place of the evaluation (laboratory vs. in situ field testing)

8.2.1 Focus Interview/Enactment/Observation Study

One of the easiest and most straightforward evaluation methods is to simply *interview* the actual/potential users and *observe their interaction behavior*, either with the finished product or through a simulated run. The interview can be conducted in a simple question-and-answer form, and can involve an actual usage of the given system/interface. Depending on the stage of the development at which the evaluation takes place, the application or interface may not be ready for such a full-fledged test drive. Thus, a simple paper/digital mock-up may be used so that a particular usage scenario may be enacted for use during a subsequent interview (Figure 8.4). While mock-ups provide a tangible product and thus an improved feel for the system/interface (vs. a mere rough paper sketch) at an early stage of the development, important interactive features may not have been implemented as yet. In this case, a *Wizard of Oz* [7] type of testing is often employed, where a human administrator fakes the system response "behind the curtain." User-interaction behaviors during the test trials or simulation runs are recorded or videotaped for more detailed postanalysis.

The interview is often focused on particular user groups (e.g., elderly) or on the features of the system/interface (e.g., information layout) to

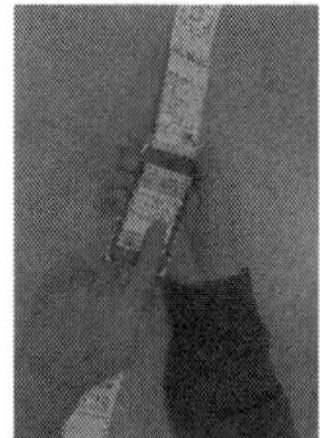

Figure 8.4 Interviewing a subject upon simulating the usage of the interface with a mock-up.

Figure 8.5 A cognitive walkthrough with the interviewer.

save time. One particular interviewing technique is called the *cognitive walkthrough* in which the subject (or expert) is asked to "speak aloud" his thought process (Figure 8.5). In this case, the technique is focused on identifying any gap between the interaction model of the system and that of the user. We can deduce that cognitive walkthroughs are fit for evaluation at a relatively earlier stage of design, namely interaction modeling or interface selection (vs. specific interface design). Another notable variation of the actual usage-based testing is the "Can you break this?" type of testing in which the subject is given the mission to explicitly expose interface problems, e.g., by demonstrating interface flaws and interface-design-related bugs.

Note that the interview/simulation method, due to its simplicity, can be used not only for evaluation, but also for interaction modeling and exploration of alternatives at the early design stage. In Chapters 3 and 4, we have already seen design tools such as storyboards, wireframing, and GOMS (Goals, Operators, Methods, and Selection), which can be used in conjunction with users or experts for simultaneous analysis and design. The user interviewing/observation technique, being somewhat free-form, is easy to administer but is not structured to be comprehensive. Table 8.1 summarizes the characteristics of the interview/simulation/observation approach.

Table 8.1 Summary: Interview, Usage, and Observation Method

Evaluators/size	Actual users/medium sized (10–15)		
Type of evaluators	Focused (e.g., by expertise, age group, gender)		
Formality	Usually informal (not controlled experiment)		
Timing and objectives	STAGE	OBJECTIVE	ENACTMENT METHOD
	Early	Interaction model and flow	Mock-up/ Wizard of Oz
	Middle	Interface selection	Mock-up/ Wizard of Oz Partial simulation
	Late/after	Interface design issues (look and feel such as aesthetics, color, contrast, font size, icon location, labeling, layout, etc.)	Simulation Actual system

Note: Free form is easy to administer, but it is not structured or comprehensive.

8.2.2 Expert Heuristic Evaluation

Expert heuristic evaluation is very similar to the interview method. The difference is that the evaluators are HCI experts and that the analysis is carried out against a preprepared HCI guideline, hence the term *heuristics*. For instance, the guideline can be general or more specific (Chapter 2) with respect to application genre (e.g., for games), cognitive/ergonomic load, corporate UI design style (e.g., Android™ UI guideline), etc. The directions or particular themes of the heuristics are chosen by the underwriter. The following lists Nielsen's ten general UI heuristics [8]. Note that these guidelines are almost the same as the general principles/guidelines introduced in Chapters 1 and 2 and used for interaction/interface design.

1. *Visibility of system status*: The system should always keep users informed about what is going on, through appropriate feedback within reasonable time.
2. *Match between system and the real world*: The system should speak the users' language, with words, phrases, and concepts familiar to the user, rather than system-oriented terms. Follow real-world conventions, making information appear in a natural and logical order.
3. *User control and freedom*: Users often choose system functions by mistake and will need a clearly marked "emergency exit" to leave the unwanted state without having to go through an extended dialogue. Support *undo* and *redo*.

4. *Consistency and standards*: Users should not have to wonder whether different words, situations, or actions mean the same thing. Follow platform conventions.
5. *Error prevention*: Even better than good error messages is a careful design that prevents a problem from occurring in the first place. Either eliminate error-prone conditions or check for them and present users with a confirmation option before they commit to the action.
6. *Recognition rather than recall*: Minimize the user's memory load by making objects, actions, and options visible. The user should not have to remember information from one part of the dialogue to another. Instructions for use of the system should be visible or easily retrievable whenever appropriate.
7. *Flexibility and efficiency of use*: Accelerators—unseen by the novice user—may often speed up the interaction for the expert user such that the system can cater to both inexperienced and experienced users. Allow users to tailor frequent actions.
8. *Aesthetic and minimalist design*: Dialogues should not contain information that is irrelevant or rarely needed. Every extra unit of information in a dialogue competes with the relevant units of information and diminishes their relative visibility.
9. *Help users recognize, diagnose, and recover from errors*: Error messages should be expressed in plain language (no error codes), precisely indicate the problem, and constructively suggest a solution.
10. *Help and documentation*: Even though it is better if the system can be used without documentation, it may be necessary to provide help and documentation. Any such information should be easy to search, be focused on the user's task, list concrete steps to be carried out, and not be too large.

In the far left and middle columns of Table 8.2, we show evaluation heuristics specifically derived for evaluating the initial design of No Sheets (done in Chapter 4). The heuristics were derived by the developer who identified, among very many possibilities, the more important principles and guidelines to follow for this particular application. The right column shows partial results of applying these evaluation heuristics. In

Table 8.2 Evaluation Heuristics Derived Specifically for Evaluating "No Sheets" and Its Application Results

HEURISTIC	SPECIFICS (EXAMPLES)	EVALUATION RESULTS (PARTIAL)
System status	Does the user understand what is going on as the song is played (e.g., part of the song is being played, current operation)?	While playing, the tempo and whether it is being played, fast-forwarded, or reviewed, is not clearly shown.
Display layout	Is the information laid out and positioned properly (e.g., chords, beat, lyrics)? Is the color-coding and icon design proper for fast recognition?	The colors are too raw (tiring to the eyes). Landscape mode is preferred (vs. portrait). The icon designs for fast-forward and review are not familiar.
Interaction/ Contents model	Are all the essential functionalities available for this application? Are the necessary functions accessible and is information displayed at different interaction points?	Tempo control, fast-forward, and review are not possible during play. Information per measures is needed.
Ergonomic consideration/User characteristics/ Operating environment	Assess readability, color contrast, and GUI object size. Also assess if easily operable in a typical/ various usage situation (for piano, guitar, etc.)	A better color contrast is needed between different types of information. Provision is needed for long lyrics. Landscape mode is more desirable.
Input/Output method	Assess interface methods: conveying the beat (beat number, sound), setting the tempo, selecting the song, etc.	Beat sound is too high pitched. Suggest dragging for fast-forward and review functions.
Consistency/ Standards	Evaluate consistency with actual sheet music and Android design guideline.	A more common choice or design of icons is needed.
Prevention of errors	Is the interaction modeled or designed such that it minimizes error? Is it possible to easily undo?	Explicitly deactivate the play button when there is no song selected.
Aesthetics	Evaluate simplicity and overall attractiveness.	Mostly simple except for using too much primary colors.
Help	Is there sufficient help and guides for the beginner?	Need more detailed guide and introduction.

this way, the evaluation was carried out efficiently by a third-party HCI expert by paying particular attention to those heuristics.

The expert heuristic evaluation is one of the most popular methods of UI evaluation because it is quick and dirty and relatively cost effective (Table 8.3). Only a few (typically three to five) UI and domain experts are typically brought in to evaluate the UI implementation in

Table 8.3 Summary of the Expert Review Method

Evaluators/size	HCI experts/small sized (3–5)		
Type of evaluators	Focused (experts on application-specific HCI rules, corporate-specific design style, user ergonomics, etc.), interface consistency		
Formality	Usually informal (not controlled experiment)		
Timing and objectives	STAGE	OBJECTIVE	ENACTMENT METHOD
	Middle	Interface selection	Scenarios Storyboards Interaction model
	Late/after	Interface design issues (look and feel such as aesthetics, color, contrast, font size, icon location, labeling, layout, etc.)	Simulation Actual system

Note: Easy and quick, but prior heuristics are assumed to exist, and no actual user feedback is reflected.

the late stage of the development or even against a finished product. The disadvantage of the expert review is that the feedback from the user is absent, as the HCI expert may not understand the needs of the actual users. On the other hand, the small size of the evaluator pool is compensated by the expertise of the participants.

8.2.3 Measurement

In contrast to interviews and observation, measurement methods attempt to indirectly quantify the goodness of the interaction/interface design with a *score* through representative task performance (quantitative) or quantified answers from carefully prepared subjective surveys (qualitative).

Typical indicators for quantitative task performance are the task completion time, score (or amount of task performance in unit time), and errors (produced in unit time). For example, for a mobile game, a representative task might be to "invoke the given game, log in, and reach the main screen." Another example task, for No Sheets, would be to "invoke the application, load the music file, and set the tempo" (Figure 8.6). Task-performance measurement is only meaningful when compared to the nominal/reference case. Thus, two measurements must be made between the nominal and the new design, and statistical analysis is then applied to derive any meaningful and significant differences between the two measurements (Figure 8.7). To

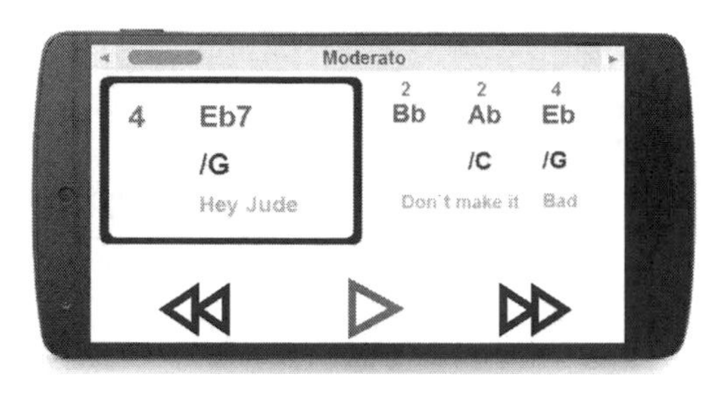

Figure 8.6 The initial (left) and redesigned (right) "play" activity/layer for No Sheets: The new design after evaluation uses a landscape mode and fewer primary colors. The icons for fast-forward and review are changed to the conventional style, and the current tempo is shown on top.

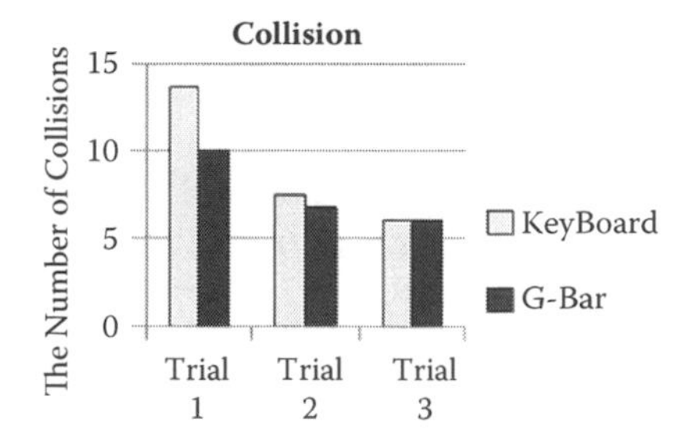

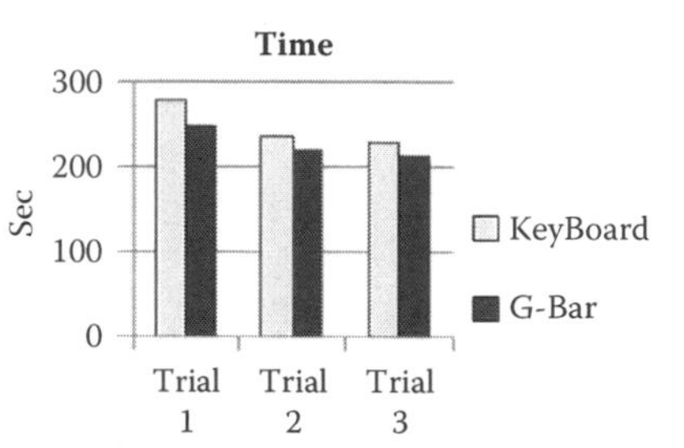

Figure 8.7 A case of a task-performance measurement: (1) nominal: a game interface using a keyboard, and (2) new: a game interface using a new controller. Task completion time for navigating a maze is measured using the respective interfaces and then compared to indirectly assess the ease of interaction.

minimize bias or variation, it is generally accepted that it is feasible to gather a sufficiently homogeneous yet relatively small subject pool for physical/cognitive task-performance measurement.

On the other hand, numerical scores can also be obtained from surveys. Surveys are used because many aspects of usability or user experience are based on user perception, which is not directly measurable. However, answers to user-perception qualities are highly variable and much more susceptible to bias by the users' intrinsic backgrounds. A few provisions can be made to reduce such biases, for example by using a large number of subjects (e.g., more than 30 people), using an odd-leveled (5 or 7) answer scale (also known as the Likert scale [9] so that there is always a middle-level answer, and carefully wording and explaining the survey questions for clarity and understanding (more guidelines in Table 8.4). Even though the result of the survey is a numerical score, the nature of the measurement is still qualitative because survey questions usually deal with user-perception qualities. Similarly to the task-performance case, a comparative survey against the nominal case is recommended.

Both types of measurement experiments can optionally be run over a long period of time, especially when memory performance and familiarity with the task is involved. For instance, to assess the ease of learning an interface, the task performance can be measured over weeks to see how quickly the user recalls how to operate the interface and produce higher performance.

Table 8.4 Guidelines for a Good Survey

Minimize the number of questions	Too many questions results in fatigue and hence unreliable responses.
Use an odd-level scale of five or seven (or Likert Scale)	Research has shown odd answer levels with mid value with five or seven levels produces the best results.
Use consistent polarity	Negative responses correspond to Level 1 and positive to Level 7 and consistently so throughout the survey.
Make questions compact and understandable	Questions should be clear and easy to understand. If difficult to convey the meaning of the question in compact form, the administrator should verbally explain.
Give subjects compensation	Without compensation, subjects will not do their best or perform the given task reliably.
Categorize the questions	For easier understanding and good flow, questions of the same nature should be grouped and answered in block, e.g., answer "ease of use" related questions, then "ease of learning," and so on.

Another variation is with the place of the evaluation. When testing with the finished product, it is best to conduct the usage test at the actual place of usage, outside the laboratory (e.g., at the office, at home, on the street, etc.). However, as expected, it is often very difficult to conduct the measurement or testing at the actual place of interaction. Even when it is possible, there are many uncontrollable factors that might affect the outcome of the testing (e.g., having to test in front of other people). To isolate and prevent these possible biases, the testing is often conducted in a laboratory setting as well, with a carefully selected pool of homogeneous subjects.

With the advent of smartphones and their ubiquity, in situ field testing is gaining great popularity [10]. Applications can collect user interaction information in the background upon particular interaction events, and this information can then be analyzed in a batch process. While the same danger exists with respect to the environmental biases, these can be often mitigated by the high number of subjects (e.g., users of smartphones and apps). Some research has shown that there is very little difference in the analysis/evaluation results between the controlled laboratory studies and the in situ field studies [11]. However, this result depends on the nature of the applications (especially those for which typical usage situations cannot easily be re-created in the laboratory) [12].

In fact, in addition to the need to carefully construct the survey, measurement experiments require meticulous operational logistics to be as fair and bias free as possible, starting from the recruitment, screening, and pretraining of the subjects, compensation for and obtaining the consent of the subjects, choosing the right independent and dependent variables, and applying the right statistical analysis methods to the resulting data. The details of such design of experiments (DEX) are beyond the scope of this book, and we refer you to the related literature. Despite the higher reliability of the evaluation results, a significant amount of effort is needed to prepare and administer the measurement interface evaluation method (Table 8.5).

8.2.4 Safety and Ethics in Evaluation

Most HCI evaluation involves simple interviews and or carrying out simple tasks using paper mock-ups, simulation systems, or prototypes.

Table 8.5 Summary of the Measurement Method

Evaluators/sample size	Potential or typical users/medium to large size (10 to 50 or more)		
Type of evaluators	Balanced and homogeneous pool of subjects (users of the system—gender, age, educational background, relevant skills, etc.)		
Formality	Can be a formally controlled experiment or an informal assessment		
Place	Laboratory or in situ field		
Timing and objectives	STAGE	OBJECTIVE	ENACTMENT METHOD
	Late/after	Interface design issues (look and feel, such as aesthetics, color, contrast, font size, icon location, labeling, layout, etc.)	Simulation Actual system

Note: More reliable results, but generally time consuming to prepare and conduct the process.

Thus, safety problems rarely occur. However, precautions are still needed. For example, even interviews can become long and time consuming, causing the subject to feel much fatigue. Some seemingly harmless tasks may bring about unexpected harmful effects, both physically and mentally. Therefore, evaluations must be conducted on volunteers who have signed consent forms. Even with signed consents, the subjects have the right to discontinue the evaluation task at any time. The purpose and the procedure should be sufficiently explained and made understood to the subjects prior to any experiments. Many organizations run what is called the Institutional Review Board (IRB), which reviews the proposed evaluative experiments to ascertain safety and the rights of the subjects. It is best to consult or obtain permission from the IRB when there is even a small doubt of some kind of effect to the subjects during the experiments.

8.3 Summary

We have looked at various methods for evaluating the interface at different stages in the development process. As already emphasized, even though all the required provisions and knowledge may have been put to use to create the initial versions of the UI, many compromises may be made during the actual implementation, resulting in a product somewhat different from what was originally intended at the design stage. It is also quite possible that during the course of the development, the requirements simply change. This is why the explicit evaluation step is a must and, in fact, the whole design-implement-evaluate cycle must ideally be repeated at least a few times until a stable result is obtained.

References

1. Wikipedia. 2014. Usability. http://en.wikipedia.org/wiki/Usability.
2. Hart, Sandra G., Steve Land, and E. Lowell. 1988. Development of NASA-TLX (Task Load Index): Results of empirical and theoretical research. *Human Mental Workload* 1 (3): 139–83.
3. NASA. 2013. NASA Task Load Index. http://humansystems.arc.nasa.gov/groups/tlx/downloads/TLXScale.pdf.
4. ISO. 2009. Ergonomics of human system interaction—Part 210: Human-centred design for interactive systems. ISO DIS 9241-210:2010. Geneva, Switzerland: International Organization for Standardization.
5. Bevan, N. 2008. UX, Usability and ISO standards. Paper presented at Values, Value and Worth workshop, CHI 2008, Florence, Italy. http://www.cs.tut.fi/ihte/CHI08_workshop/papers/Bevan_UXEM_CHI08_06April08.pdf.
6. Lewis, James R. 1995. IBM computer usability satisfaction questionnaires: Psychometric evaluation and instructions for use. *International Journal of Human–Computer Interaction* 7 (1): 57–78.
7. Wikipedia. 2013. Wizard of Oz experiment. http://en.wikipedia.org/wiki/Wizard_of_Oz_experiment.
8. Nielsen, Jakob. 1994. Enhancing the explanatory power of usability heuristics. In *Proceedings of the SIGCHI conference on human factors in computing systems*, 152–58. New York: ACM Press.
9. Likert, Rensis. 1932. A technique for the measurement of attitudes. *Archives of Psychology* 22 (140): 1–55.
10. Rowley, D. E. 1994. Usability testing in the field: Bringing the laboratory to the user. In *Proceedings of the SIGCHI conference on human factors in computing systems*, 252–57. New York: ACM Press.
11. Kaikkonen, A., T. Kallio, A. Kekalainen, A. Kankainen, and M. Cankar. 2005. Usability testing of mobile applications: A comparison between laboratory and field testing. *Journal of Usability Studies* 1 (1): 4–16.
12. Kjeldskov, J., M. B. Skov, B. S. Als, and R. T. Høegh. 2004. Is it worth the hassle? Exploring the added value of evaluating the usability of context-aware mobile systems in the field. *Lecture Notes in Computing Science* 3160:61–73.

9
Future of HCI

Human–computer interaction (HCI) has contributed much to the advancement of computing and its spread into our everyday living. The prevalent type of interface up to the late twentieth century was the so-called WIMP (windows, icon, mouse, pointer) and graphical user interface (GUI) for the stationary desktop computing environment. This was a huge improvement over its predecessor, the keyboard-input command-oriented interface. Much innovation has been made on the two-dimensional (2-D)-oriented desktop interface since it was first introduced in the early 1980s. These include ergonomic mouse and keyboard design, hypertext and web interface, user interface toolkits, extension of the Fitts's law, interaction modeling, and evaluation methodologies. If you look more closely, the innovation in HCI has always followed or been accompanied by an advancement of the hardware and software platforms. Even though the original concept of the mouse and graphical user interface was actually devised in the late 1960s by Doug Engelbart, it was not until the early 1980s that the hardware and software technology (not to mention the possibility of personal computing as hardware prices became much more affordable) was mature enough to accommodate the use of a mouse and the GUI (Figure 9.1).

This line of thought can give us a good glimpse into the future of HCI based on the fast-changing trends in computing platforms. Here are four major new computing platforms that have emerged in the past 10 years:

- *Mobile and handheld platform*: (exemplified by the smartphones) which we can carry around to compute and communicate
- *Ubiquitous platform*: in which everyday objects are embedded with interactive computing/networking devices and services
- *Natural and immersive computing/sensing/display platform*: that provides near-realistic services and experiences

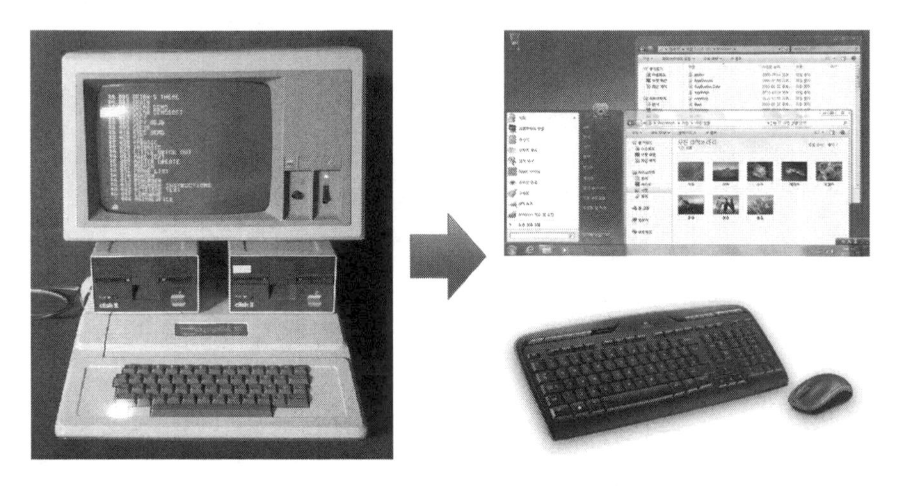

Figure 9.1 Keyboard-input command-oriented interface compared to the WIMP- and GUI-based interfaces (1980–1990).

- *Cloud computing platform*: that provides high-quality interactive services (based on its heavy-duty ultraserver-level computing power) with real-time response (based on the fast network service)

In the case of the cloud computing platform, the typical user will not interact directly with the system where the application resides (somewhere in the cloud), but through the client computer or device, such as the everyday desktop computers and mobile devices. Despite the tremendous growth in the computing power of desktop and even mobile units, these stand-alone machines are not usually sufficient for such high-end interactive and intelligent services as image recognition, language understanding, context-based reasoning, and agentlike behavior. Note that these so-called client devices (for the cloud) are becoming increasingly richer in their sensing, display, and network capabilities. In essence, the cloud is taking up the role of the Model and the client View/Controller, where there can be many View/Controllers for different types of clients (e.g., desktops, pads, smartphones). This can be viewed as a way to improve the user experience (UX) by providing high-quality services in real time and having specialized interaction clients focused on usability that are easily deployed (due to their lightness and mobility). For such an envisioned future, it will be necessary to develop middleware solutions that will manage the seamless connection between the Model and one of many possible client View/Controllers.

On the other hand, in terms of hardware, we expect that the mobile and ubiquitous platforms will accelerate and further drive the integration of embedded computers, sensors, and sensor networks into everyday objects (as is the goal of the Internet of Things*). Touch technology is the main interaction mode for mobile and embedded/ubiquitous computers and devices, and this technology will become more refined as it evolves to accommodate multitouch, proximity touch (hovering), and touch-haptic feedback.

The interaction styles of the mobile/embedded vs. natural/realistic/immersive interfaces can be understood in terms of people's natural dichotomous desires: one for simple and fast operations in a dynamic environment and the other for a rich and experiential interaction in a more stable, relaxed environment. These two desires are in tune with the lifestyle in the coming ages as we become more affluent and culturally richer. Virtual and mixed reality, multimodal interfaces are in the forefront of the experiential interaction technologies.

Finally, as we have indicated in Chapter 8, in pursuit of the mystical UX, more interfaces are becoming affective, calling out to our emotional side. It is difficult to define what constitutes an affective interface. It could be something as simple as an emphasis on the aesthetics. It could mean personalization and adaptation catering to the user's unique and changing tastes and needs. However, the latter still remains a technological challenge, as it requires intelligent sensing and robust recognition of contexts, user emotions, and subtle intent—a very difficult task even for humans themselves. However, the machine–intelligence technology continues to make almost unimaginable leaps, as demonstrated by the recent IBM Watson computer that has beaten a human champion in the quiz-show contest *Jeopardy* [1]. In the following sections, we take a closer look at these promising HCI technologies, many of which are in an active stage of research (Figure 9.2).

9.1 Non-WIMP/Natural/Multimodal Interfaces

In Chapter 4, we studied the process of HCI design, and after considering various requirements, user characteristics, and operating

* Internet of Things (IoT) refers to a concept where everyday objects are embedded with Internet capability so that the object information can be accessed and controlled.

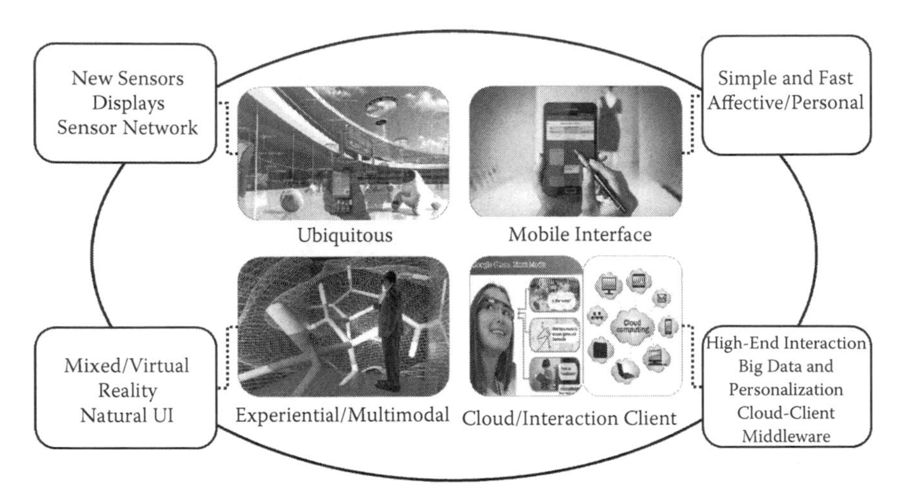

Figure 9.2 Four emerging computing platforms and associated HCI technologies to pay attention to in the next 10 years: high-quality cloud service and ubiquitous and mobile interaction clients, experiential and natural user interfaces.

constraints, we found that the available interface choices were limited, as we had to consolidate different possible solutions according to the restrictions imposed by the practical limitations of the computing platforms available today (e.g., WIMP for desktop and touch-based for smartphones). However, the future will bring the development of many different computing platforms, and we are bound to have more choices, including non-WIMP-type of interfaces that will provide more natural and multimodal interfaces. One of the main reasons these non-WIMP interfaces have not yet made it into the mainstream, despite the apparent need, is the lack of robustness and accuracy, or from another perspective, the relatively large amount of computation required to achieve them. However, the situation is changing due to continued technological innovation and the emergence of the cloud computing infrastructure. In the light of this trend, we now review and assess the future of these HCI technologies one by one, including language understanding, gesture recognition, image recognition, and multimodal interaction.

9.1.1 Language Understanding

The talking computer interface is undoubtedly the holy grail of HCI. Language understanding can be largely divided into two processes. The first is recognizing the individual words, and the second is making sense out of the sentence, which is composed of a sequence of recognized words (usually known as natural language understanding). Surely word recognition (which could be spoken, written, or printed) is the prerequisite to the sentence understanding. (Here we focus only on the spoken word or voice recognition.) Voice-recognition performance and its practicality are dependent on the target number of words to be recognized, the number of speakers, the level of the noise in the usage environment, and the need for any special devices (e.g., noise-canceling microphone). The current state of the art seems to be (a) over 95% recognition rate (individual words) for (b) at least millions of words and more than 30 languages (c) in real time (through the high-performance cloud) (d) without speaker-specific training (by age, gender, dialects) (e) in a midlevel noisy environment (e.g., office with ambient noise of around 30–40 dB) and (f) with the words spoken relatively closely to cheap noise-canceling microphones or software [2]. Such a state of the art seems to be quite sufficient for a more widespread presence of voice recognition in our current lives, but it is not so except for special situations of disability support or for operating constraints in which both hands are occupied. One main reason seems to be that the users are less tolerant to the 2%–3% of incorrect recognition performance, even though humans themselves do not possess 100% word-recognition capability. Another reason might have to do with the segmentation problem. Often, voice recognition requires a mode during which the input is given in an explicit way, because otherwise it is quite difficult to separate and segregate the actual voice input from the rest (noise, normal conversation) within the stream of voice. The entrance into this mode will typically involve simple additional actions, such as a button push/release. However, users take this to be a significant nuisance in usage.

One way to overcome this problem is to rely more on multimodality. To eliminate the segmentation problem, the voice input can be accompanied by certain other modal actions, such as a gesture/posture and lip movements within a given context so that it is distinguished from noise, other people's speech, or unrelated conversation. We will discuss this multimodal integration in Section 9.1.4.

While isolated word recognition is approaching a nearly 100% accuracy rate, when trying to understand a whole sentence, individual words need to be recognized from a continuous stream of words. By a simple calculation, we can easily see that recognizing a sentence with five words, with each word having a recognition rate of 90%, will yield a success rate of only $0.9^5 = 0.59$ success rate. Add the problem of extracting the meaning of the whole sentence, and now we have an even lower success rate in the correct natural language understanding.

Despite these difficulties, due to its huge potential, great efforts are continually being made to improve the situation. The recent cases of Apple® Siri [3] and IBM® Watson [1] illustrate the bright future we have with regard to voice/language understanding. Apple Siri understands continuously spoken words and understands them with higher accuracy by incorporating the contextual knowledge of mobile device usage. IBM Watson showcased a very fast understanding of the questions asked in natural language in its bout with the human champion (however, the questions were asked in text, not in voice). While the computer used in the quiz contest was a near supercomputer-level server, IBM is developing a more compact and lighter version specialized to a specific and practical domain such as medical expert systems and IPTV (Internet protocol television) interaction [4]. AT&T provides a similar voice/language-understanding architecture for mobile phone usage, as shown in Figure 9.3.

9.1.2 Gestures

Gestures play a very important role in human communication, in many cases unknowingly. Gestures alone can convey meaning, or they can function in a supplemental role in other modes of communication. Consequently, the objective of incorporating gestures into human-computer interaction is a natural outcome. While there may be many

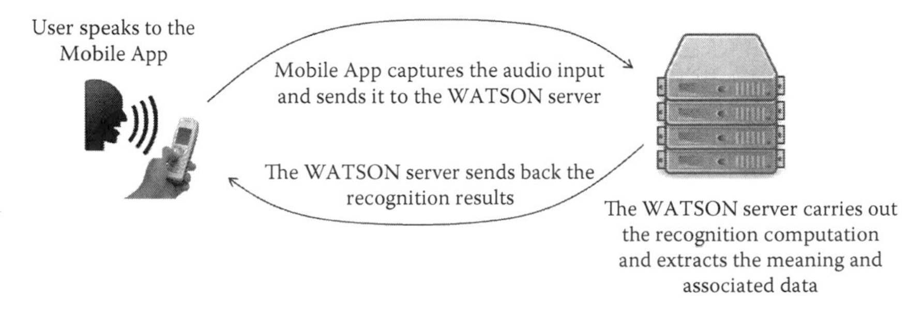

Figure 9.3 Voice/language-understanding service by the AT&T Watson cloud engine. (From AT&T Labs Research, http://www.research.att.com/articles/featured_stories/2012_07/201207_WATSON_API_announce.html?fbid=RexEym_weSd.)

different types of gestures either from the human's perspective (e.g., supplementary pointing vs. symbolic) or from the technological viewpoint (e.g., static posture vs. moving hand gestures), perhaps the most representative one is the movement of the hand(s). Hands/arms are used often for deictic gestures (e.g., pointing) in verbal communication. For the hearing-impaired, the hands are used to express sign language.

To interpret gestures, the gesture itself, whether it is a static posture or involves movement of limb(s), must be captured over time. This is generally called *motion tracking* and can involve a variety of sensors that are targeted for many different body parts. Here we illustrate the state of the art by first looking at the problem of hand tracking. Good examples of two-dimensional (2-D) hand/finger tracking are the ones using the mouse and touch screen. These technologies are quite mature and highly accurate, helped by the fact that the tracked target (hand/finger) is in direct contact with the devices. In the case of the mouse, the user has to hold the device, and this is a source of nuisance, especially if the user is to express 2-D gestures rather than just using it freely to control the position of the cursor. This explains why mouse-driven 2-D gestures have not been accepted by users, their application being limited so far to just a few games [5]. On the other hand, simple 2-D gestures on the touch screen, such as swipes and flicks, are quite popular.

With the advent of ubiquitous and embedded computing, which in many cases will not be able to offer sufficient area/space for 2-D touch input, understanding of aerial gestures in the 3-D space, which is actually closer to how humans enact gestures in real life and understand by vision, will become important. Tracking of 3-D motion of body parts or moving objects is a challenging technological task. The "inside-out" method requires the user to hold (e.g., 3-D mouse, Wiimote) or attach a sensor to the target body part or object (e.g., hand, head), with both options being perceived as being cumbersome and inconvenient (Figure 9.4). These sensors operate based on a variety of underlying mechanisms such as detecting the phase differences in electromagnetic waves, inertial dead reckoning with gyros/acceleration sensors, triangulation with ultrasonic waves, etc. The "outside-in" method requires an installation of the sensor in the environment, external to the user's body. Using the camera or depth sensors (e.g., Microsoft® Kinect) are examples of the outside-in method. Since the user is free of any devices on one's body, the movement and

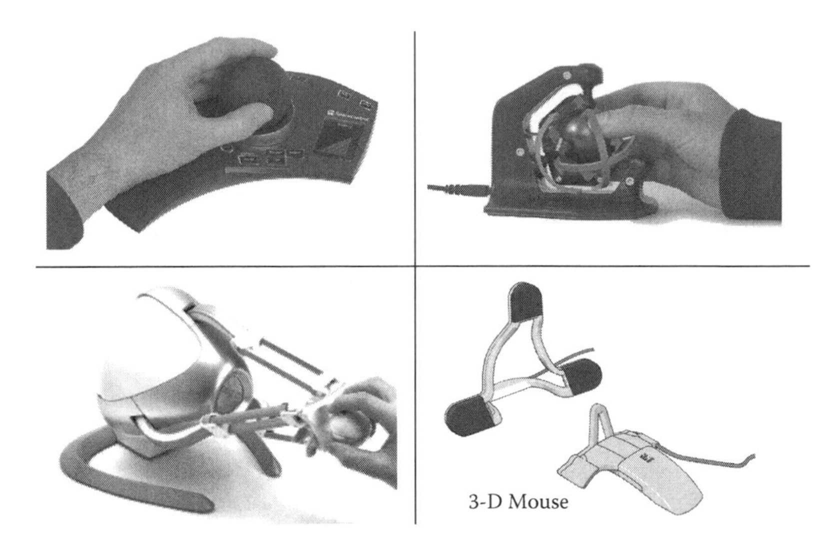

Figure 9.4 Examples of inside-out type (handheld) of sensors (3-D mouse; see 3-D Mouse and Head Tracker Technical Reference Manual, http://www.logitech.com) for 3-D motion tracking and interaction. (From SpaceControl 3D Maus Ball, http://www.spacecontrol-industries.de/14.html?&L=2Valentinheun 6D.)

gestures become and feel more natural, comfortable, and convenient. However, with the sensors being remote, the tracking accuracy is relatively lower than it is for the inside-out methods.

In recent years, camera-based tracking has become a very attractive solution because of innovations in computer-vision technologies and algorithms (e.g., improved accuracy and faster speed), lowered cost and ubiquity of the technology (virtually all smartphones, desktops, laptops, and even smart TVs are equipped with very good cameras), ever-improving processing power (e.g., CPU, GPU, multimedia processing chips), the availability of standard and free computer-vision/object-recognition/motion-tracking libraries (OpenCV* and OpenNI†), and the ease of their programming (processing language).

There are still some restrictions. For example, performance of camera-based tracking is susceptible to environmental lighting condition (Figure 9.5). For highly robust tracking, markers (e.g., passive objects that are easily and robustly detectable by computer-vision algorithms) are used, which makes the situation similar to using the inside-out

* Open Source Computer Vision (OpenCV), http://opencv.org/.

† OpenNI, the standard framework for 3-D sensing, http://www.openni.org/.

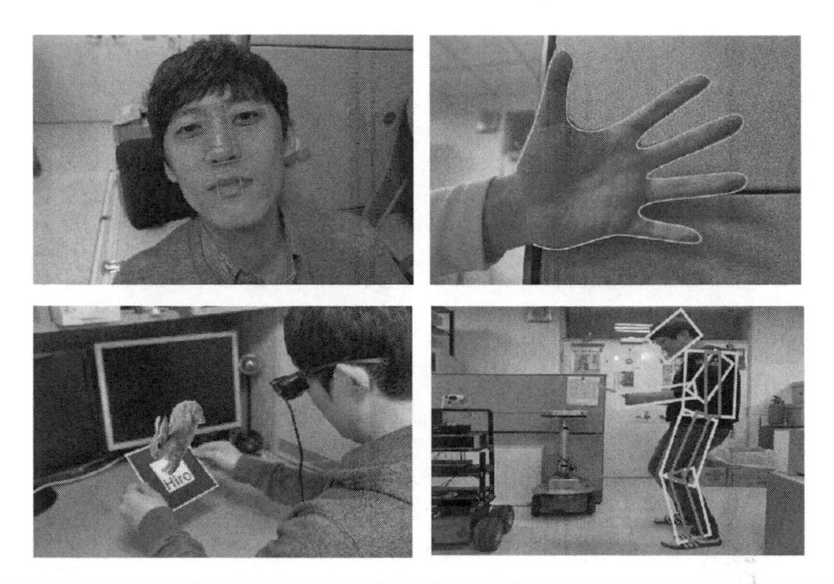

Figure 9.5 Camera-based motion-tracking examples (for face, hand, marker, and whole body).

method. Examples of markers include objects with high-contrast geometric patterns, colored objects, and infrared LEDs.

The inexpensive depth sensors introduced in the market recently have revolutionized the applicability, robustness, and practicality of the outside-in gesture and motion-based interaction. For example, the Microsoft Xbox game platform uses both a color camera and a depth sensor (originally developed by PrimeSense) and can track the whole-body skeleton motion (e.g., up to more than 10 joints) of multiple users without any devices worn on the body (Figure 9.6). It was originally intended for motion-based whole-body games, and

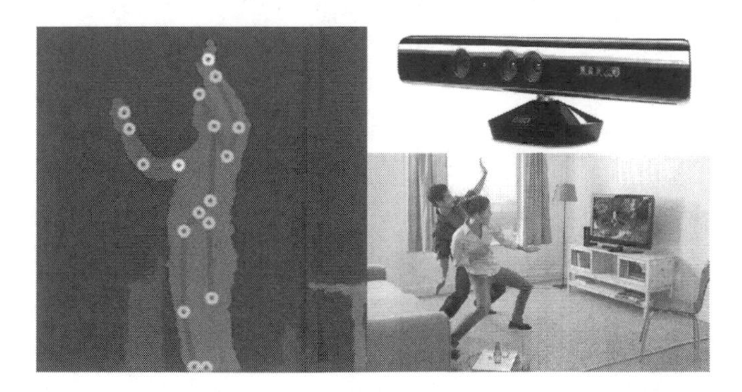

Figure 9.6 Whole-body skeletal tracking using the Kinect depth sensor (left) and its application to motion-based games (right).

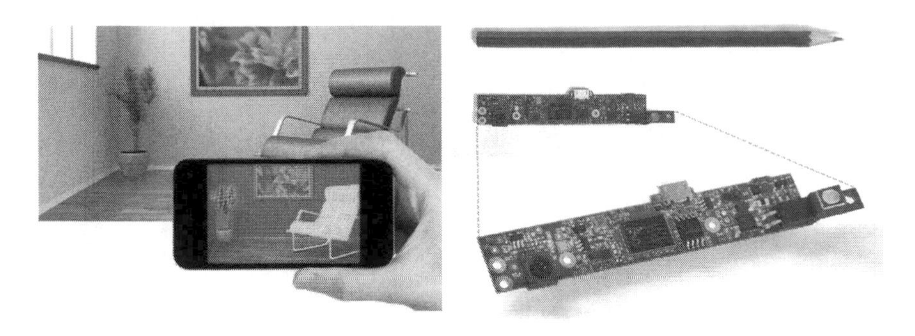

Figure 9.7 Prototype miniature depth sensor mountable on mobile devices. (From Engadget, PrimeSense demonstrates Capri 3D sensor on Nexus 10, 2013, http://www.engadget.com/2013/05/15/primesense-demonstrates-capri-3d-sensor [6].)

now its application has been extended to environment reconstruction (i.e., scanning objects in the environment to derive computer models), motion capture, and many others. The smaller, miniaturized prototype (with comparable resolution and performance) for mobile devices has already been developed [6] (Figure 9.7).

With all this said, it seems that the major hurdle has been eliminated on our road to more widespread use of motion-based interaction. Yet there still remains one more problem, which is again the same "segmentation" problem that was associated with voice recognition. Similarly, it is a difficult problem to segment the meaningful gestures out of the continuous-motion tracking data. Figure 9.8 illustrates the problem and its difficulty. Again, many current motion-gesture systems rely on operating in a particular mode (e.g., applying the gesture while pressing a button, or being in a particular state). However, this defeats the very purpose of the bare hand and truly outside-in sensing. Plus, as already stated, this additional step in the interaction, having to enter the gesture-input mode, lowers the usability dramatically. Innovative algorithms such as those based on

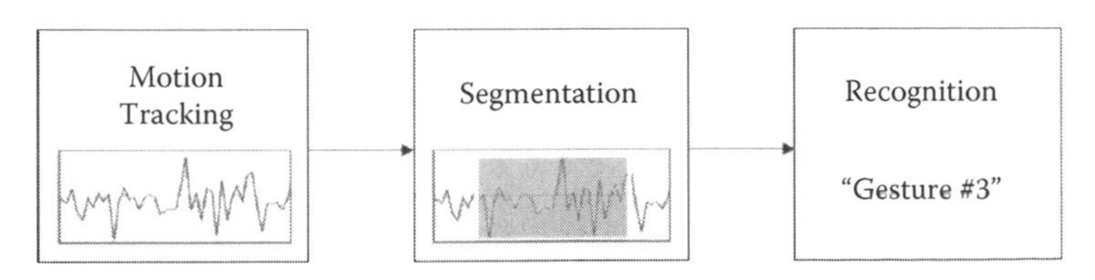

Figure 9.8 Three major steps in gesture recognition: (1) motion tracking, (2) segmentation (using the monitoring through the "sliding window" into the tracking data stream), and (3) recognition given the tracking data segment.

the concept of "sliding windows" (continuously monitoring a fixed or variable length of motion stream for the existence of a meaningful gesture) may be able to solve this problem.

The segmentation problem is more challenging for gesture recognition because, in the case of voice recognition, the background noise may be low and the detectable spoken inputs intermittent, meaning that the voice-recognition mode can be automatically activated by sound detection (e.g., sound intensity is greater than some threshold). Touch gesture is the same. In most cases, it is natural to expect touches only when a command is actually needed. Thus a touch simply signals the start of the gesture input mode. As for 3-D motion gestures, users usually continually move, and only part of it may be gestural commands that need to be extracted. Again, as we have indicated, multimodal interaction can partly solve this problem. Finally, in terms of usage, while motion-based interaction may be experiential and realistic, one must remember that it is easily tiring.

So far, we have mostly explained our point using hand or bodily motion and discussed potential difficulties in its detection and recognition. Another special case of using gestures is that of using fingers. Due to the current resolution of the sensors and the relative size of fingers against the larger human body, it is not very easy to detect the subtle articulation of the fingers. Again, with the current trends in new sensor development and declining cost, this will not be such a big problem in the near future. Depth sensors specialized for finger tracking are already appearing in the market (e.g., Leap Motion [7]). In fact, finger tracking used to be handled in the inside-out fashion by employing glove-type sensors. Wearing gloves and interacting with a computer turned out to be very cumbersome, with low usability. More importantly, regardless of the type of sensors used, it is not clear how valuable finger-based interaction might be in improving the UX. In real life, fingers are mostly used for grasping and rarely as gestures (except for the special case of sign language). Even finger-touch gestures (for touch-screen interaction) are not that many (e.g., swipe, flick, pinch). It may be possible to define many finger-based gestures once detailed finger tracking is technologically feasible, but its utility is questionable (Figure 9.9). Electromyogram (EMG) sensors are newly used to recognize motion gestures. EMG sensors can approximately detect the amount of joint movement. Figure 9.10 shows a

Figure 9.9 Finger-based interaction using the Leap Motion. (From Leap Motion, Leap Motion Controller, 2014, http://www.leapmotion.com [7]).

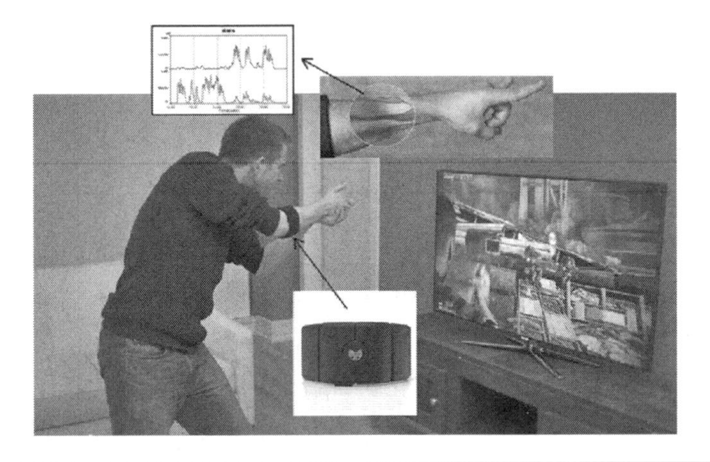

Figure 9.10 Wristband type of EMG sensor for simple gesture recognition (http://www.thalmic.com).

wristband type of EMG sensor with which a user is making a gun-triggering gesture in a first-person shooting game.

9.1.3 Image Recognition and Understanding

Image recognition or understanding is perhaps a lesser used technology in HCI, especially for rapidly paced and highly frequent interaction in which the use of mouse/touch/voice input is more common. For instance, the most typical use for face recognition might be for initial authentication (as part of a log-in procedure). Object image recognition might be used in an information search process as an alternative

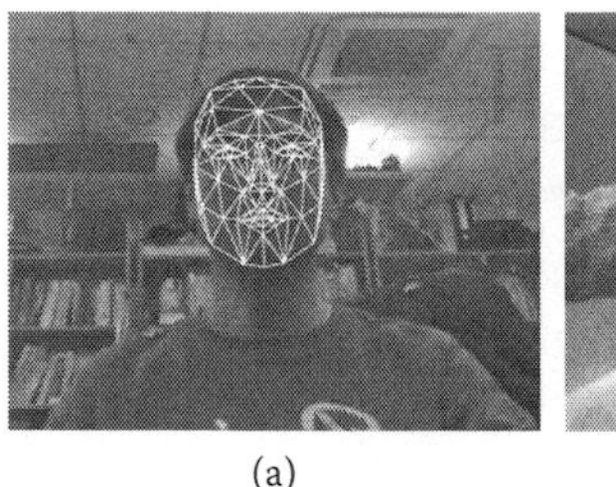

(a)

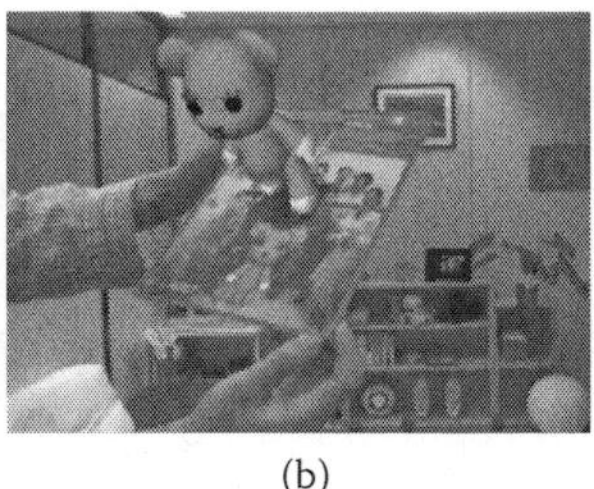

(b)

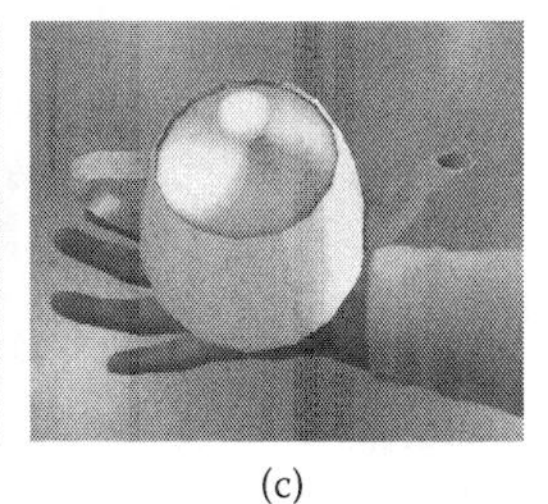

(c)

Figure 9.11 Image recognition for (a) face, (b) object/marker (Sony Smart AR, http://www.sony.net/SonyInfo/News/Press/201105/11-058E), and (c) hand and their applications for motion tracking and augmented reality.

to the usual keyword text-driven approach, e.g., when the name of the object is not known or when it happens to be more convenient to take the photo than typing in or voicing the input. Rather, the underlying technology of image recognition is more meaningful as an important part of object motion tracking (e.g., face/eye recognition for gaze tracking, human body recognition for skeleton tracking, and object/marker recognition for visual augmentation and spatial registration).

Lately, image understanding has become even more important, as the core technology for mixed and augmented reality (MAR) has attracted much interest. MAR is the technology for augmenting our environment with useful information (Figure 9.11). With the spread of smartphones equipped with high-resolution cameras, GPUs, and light and fashionable see-through projection glasses (not to mention near 2-GHz processing power), MAR has started to find its way into mainstream usage and may soon revolutionize the way we interact with everyday objects. Moreover, with the cloud infrastructure, the MAR service can become even more robust and high quality. Finally, image recognition can also assume a very important supplementary role in multimodal interaction. It can be used to extract affect properties (e.g., facial expression), disambiguation of spoken words (e.g., deictic gestures), and lip movements). See Figures 9.12 and 9.13.

9.1.4 Multimodal Interaction

Throughout this chapter, I have alluded to the need for multimodal interaction on many occasions. Even though machine recognition rates in most modalities are approaching 100% (with the help of the

Figure 9.12 Bolt's pioneering "put that there" system. The target object of interest is identified from voice and deictic gestures. (From Bolt, R.A., *Proceedings of ACM SIGGRAPH*, Association for Computing Machinery, New York, 1980, pp. 262–270 [8].)

Figure 9.13 Applying image understating to information search and augmentation on a wearable display device (Google® Glass, https://plus.google.com).

cloud–client platform), the usability is not as high as we might expect because of a variety of operational restrictions (e.g., ambient noise level, camera/sensor field of view, interaction distance, line of sight, etc.). To reiterate, this is where multimodal interaction can be of great help. In this vein, multimodal interaction has been an active field of research in academia beginning with the first pioneering system, called "put that there," developed by Bolt et al. at MIT in the early 1980s [8]. Since then, various ways of combining multiple modalities for effective interaction have been devised (Figure 9.14). Although we have already outlined them in Chapter 3, we list them again here.

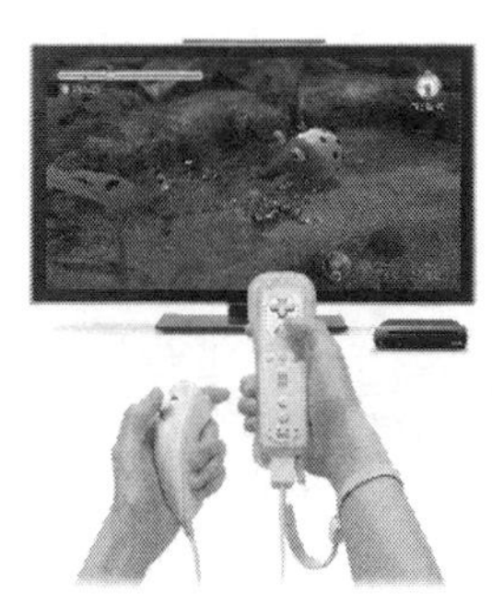

Figure 9.14 Multimodal interaction in games (left) using the buttons for setting selection, and (right) action gestures for the game play itself (Microsoft Kinect, http://www.xbox.com/en-us/kinect; Nintendo® Wii, http://www.nintendo.com/wiiu/features).

- *Composed*: In this scheme, for a set of subtasks (which together satisfy a larger task), we assign the most appropriate modality to each task. Thus each modality takes up a different role in the interaction. The "put that there" system was one such example, where the voice was used to understand the action command (verb) and the deictic gesture to identify the target object (pronoun). By "most appropriate" we assume and mean that a certain modality is most fitting and natural for a certain type of action. For instance, in a game application, it can be argued that various settings (e.g., selection of the character, weapon, sound options, etc.) can be accomplished with voice or touch interaction for the highest efficiency, while the game itself is played using action gestures for the experience. Note that *multimodal interaction* does not necessarily mean that different modal interactions occur simultaneously.
- *Alternative*: In this scheme, as the name suggests, multiple modal interaction techniques are used for the same subtask independently. The choice is made purely by user preference or by the operational situation. When dialing in a regular situation, one might use the touch interaction, while during driving, voice interaction can be used instead. This way the usability is improved by catering to the user's preferences and needs (Figure 9.15).
- *Redundant*: In the redundant scheme, many modalities are used together (simultaneously or not) for the same task (input or output). As an interaction method, it makes the act of

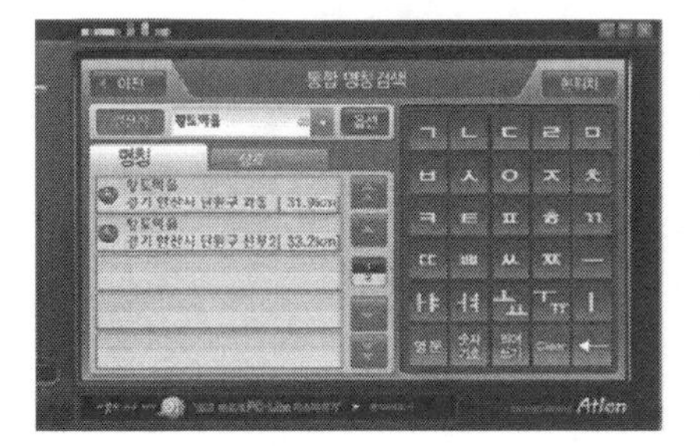

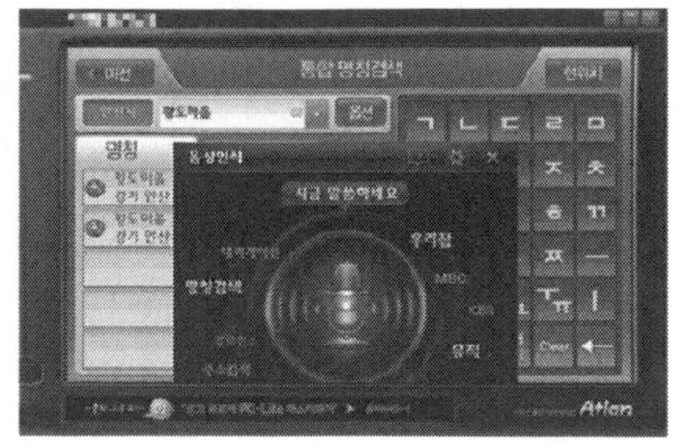

Figure 9.15 Alternative multimodal interfaces in the vehicle navigation systems (touch and voice). (From Finedrive, http://www.fine-drive.com.)

conveying the intent or information much more robust by combining those of the individual. For instance, an indication of an incoming phone call can use all three modalities: visual, aural, and tactile (vibration). With all three modalities in play, the user is less likely to miss a phone call (Figure 9.16).

Another advantage of multimodal interaction is that, to some degree, parallel interaction is possible. We often find people multitasking in different modalities, e.g., walking, listening to music, and texting. The extent of this ability is still a research question. However, it seems quite certain that for this to happen (in an effective and

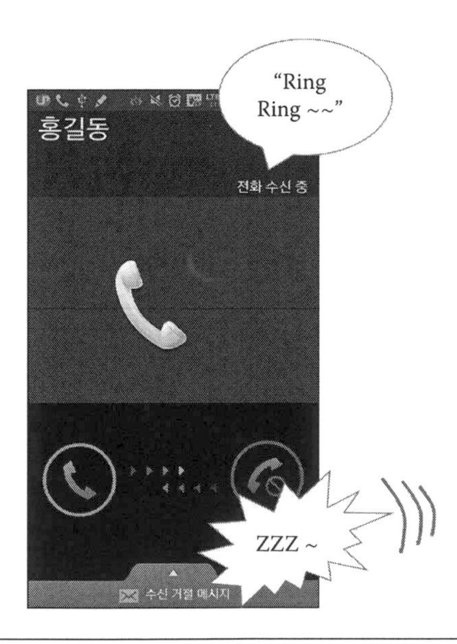

Figure 9.16 Redundant multimodal output for an incoming phone call using visual, aural, and tactile modalities.

meaningful way), the (multi)tasks must be independent of each other. If each modal interaction shares a common resource, it can be difficult to multitask concurrently (e.g., listening to music, interpreting the words, and dancing to it).

Thus, designing for multimodal interaction requires careful considerations of things like modality appropriateness (for the task), cognitive resource usage, synchronization (e.g., multiple modalities perceived as one event when temporally synchronized with a short amount of time), balance (e.g., one modality is not relatively dominating over another), and consistency (e.g., providing consistent information content between simultaneous multimodal input/output).

9.2 Mobile and Handheld Interaction

It goes without saying that smartphones have now almost replaced the PC, at least in terms of casual computing and even as a big part of business computing. As such, the importance of usability and UX for mobile and handheld interaction is even higher than ever. It is also interesting that the mobile device, as represented by the smartphones, is a focal point toward which the two notable future trends are converging: (a) multimodal interaction (with all the on-mobile sensors and displays) and (b) cloud-based services (through high-speed wireless communication).

In this context, more research is needed in the ergonomic aspects of multimodal interaction for the active (e.g., while moving), dynamic (e.g., frequently changing operating environment), and multitasking lifestyle. At least one notable trend in the mobile interaction is the simple and quick approach (vs. the rich experiential approach). It is not surprising that people would prefer the simple and quick interfaces in the midst of the modern hectic lifestyle, even for entertainment applications such as games. Many recent successful mobile games are those that are called "casual," in which a single game play session lasts only about a minute with single-touch operation and almost no learning required. On the other hand, home-based computing platforms (e.g., game consoles, smart TV, desktop), which would be used in a more relaxed atmosphere, are becoming more natural, immersive, and experiential (Figure 9.17).

As part of the cloud, and to supplement and complement the on-mobile sensors, one particular service to take note of is the sensor network

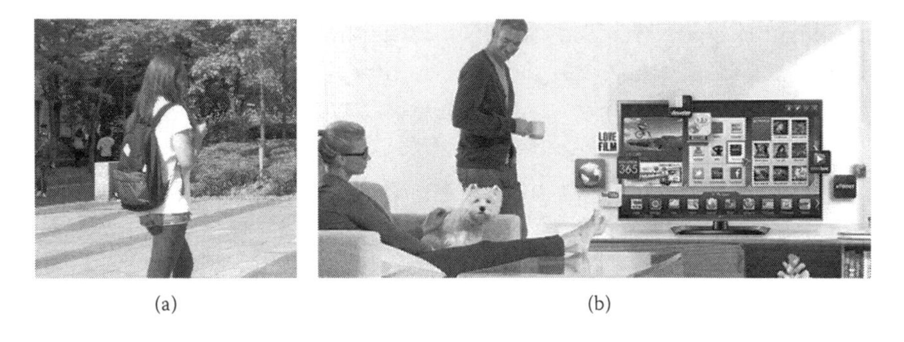

Figure 9.17 Two bipolar directions in future interaction style: (a) "simple and quick" mobile/handhelds and (b) rich and experiential stationary platforms at home. (From LG Smart TV, http://www.lg.com/tw/smart-tvs.)

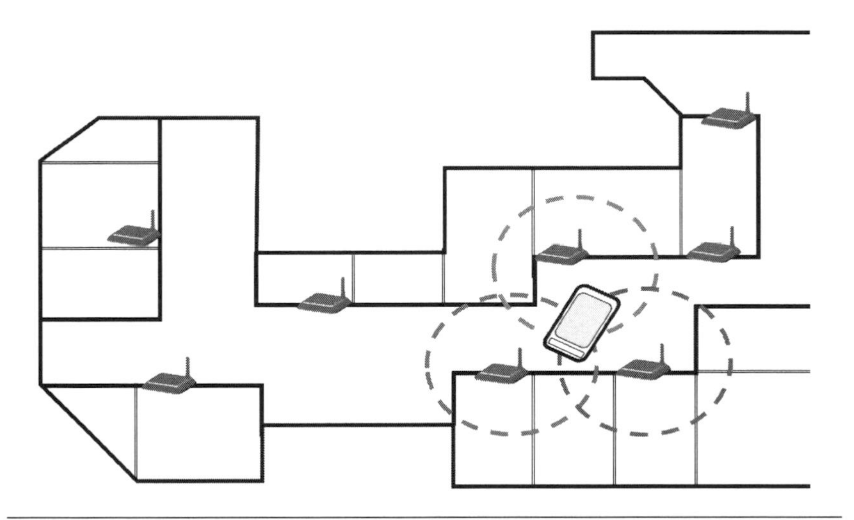

Figure 9.18 Indoor tracking of mobile devices/users using a Wifi sensor network.

service, i.e., a network of sensors in the environment collectively providing certain services mediated through the cloud. For example, sensor networks can help the mobile client infer the context of usage (e.g., location/area, lighting condition, time, number of people in the vicinity, outdoor/indoor) and provide UX at the personalized level (Figure 9.18).

9.3 High-End Cloud Service: Multimodal Client Interaction

Many interaction technologies require artificial intelligence (AI). After all, recognizing spoken words, sentences, images, and gestures are hallmarks of human intelligence. Advanced AI generally requires large databases, long off-line learning processes, and often heavy online computation (for real-time responses). High-performance servers coupled with mobile clients that handle the fast input data capture and transfer offer an attractive solution. For example, Qualcomm® Vuforia™ [5] is a cloud-based solution for image recognition that can be used for a variety of interactive services such as augmented reality and image-based search. To develop an interactive image-based service, the developer first registers images of target objects to be recognized in the server ahead of time. These input target images are trained off-line on the server so that they can be recognized well from different viewpoints at different scales and lighting conditions. The mobile application captures an arbitrary image and sends it to the server built with references to the target images of interests. The recognition computation is carried out on the server, with the results sent back to the mobile application for further processing (e.g., augmentation on the screen), all in real time (Figure 9.19).

Such a division of computational labor is reminiscent of the old time-shared computing scheme. The implication is that such a framework is readily applicable for a variety of HCI-related computations such as context-based reasoning, multimodal integration, user characteristics deduction, large-scale and multi-user tracking, usage pattern analysis, client platform adaptation, crowd-sourcing and big data gathering, environment sampling methods, etc. Figures 9.20 and 9.21 illustrate the future vision, in which a middleware installed both at the cloud and the client mediate the seamless integration between the two. For instance, the client can register itself with the cloud with information

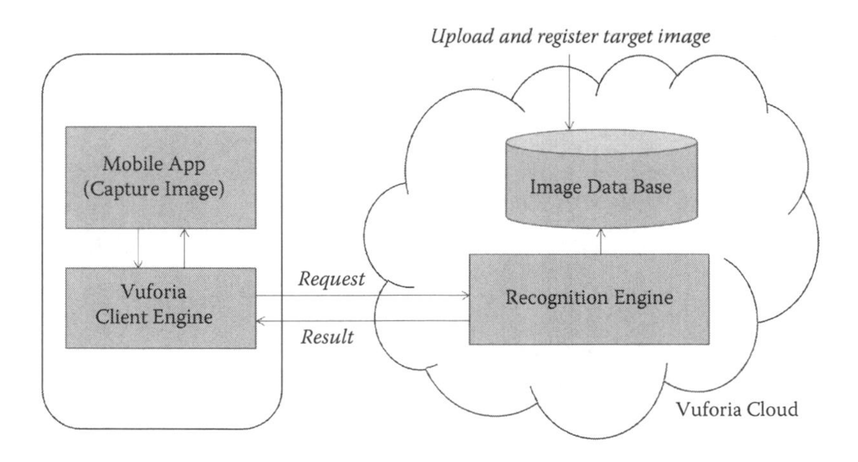

Figure 9.19 A cloud-based image-recognition service from Qualcomm Vuforia. (From Qualcomm Vuforia, https://developer.vuforia.com/resources/dev-guide/getting-started.)

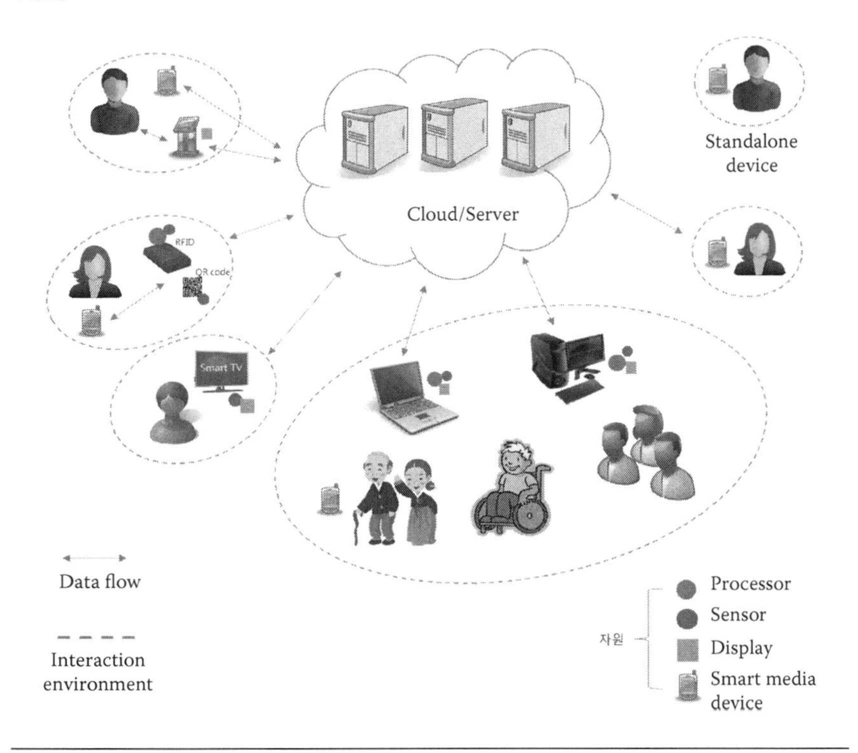

Figure 9.20 The middleware between the cloud and interaction client will enable the vision of "one application–many devices" without separate platform-specific implementations.

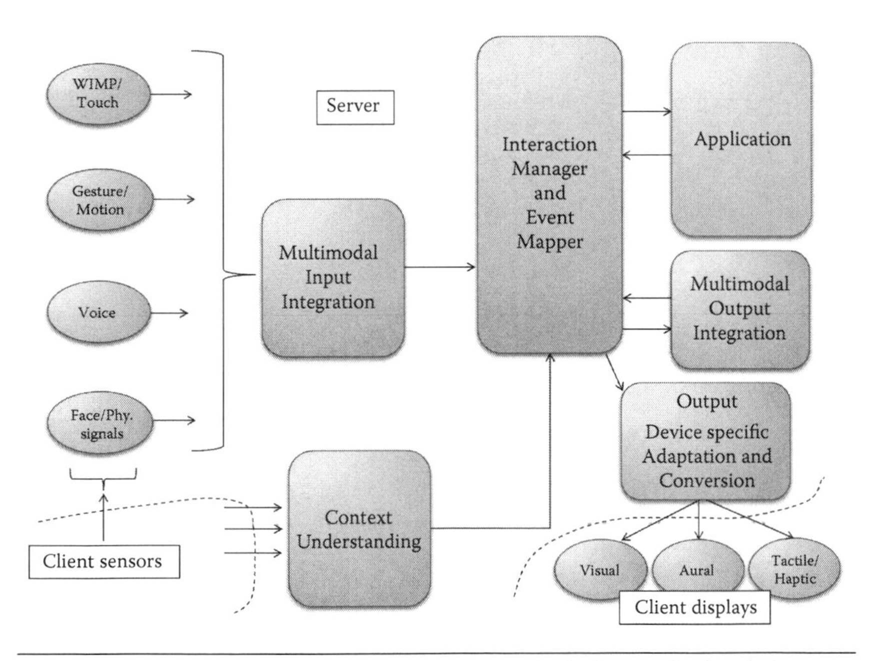

Figure 9.21 Middleware architecture for supporting the cloud–client application platform.

of its sensing and display capabilities. Interactions of the applications in the server can be described and coded only in abstract terms and communicated to the client for actual realization based on the known capabilities of the client device. This way, different models and types of devices can use the same cloud applications and services with interaction customized for users and their particular devices.

9.4 Natural/Immersive/Experiential Interaction

Today's home-computing environment is fast changing with the evolution of the television. The "smart" TVs are no different than a high-performance computer with a network connection. Moreover, smart TVs in the living room serve as the center of entertainment, and they are becoming more and more "high-fidelity," e.g., PC-level computing power, more than a 42-in. screen with UHD (ultra-high definition) resolution and stereoscopy, 5.1 surround sound, sensors (camera, depth sensor, microphone, etc.), and fiber-optic/land-line network connection. The recent successes of the Microsoft Kinect and Nintendo Wii games attest to this future trend. Thus we can even expect things like haptic sofas, living-room table computing, and simple olfactory displays. The applications will eventually extend, initially from entertainments, to immersive teleconferencing for home offices and VR (virtual reality)-based training and education (Figure 9.22).

Figure 9.22 VR-based home entertainment system. (From Xbox Kinect, http://www.xbox.com/ko-KR/Kinect/School.)

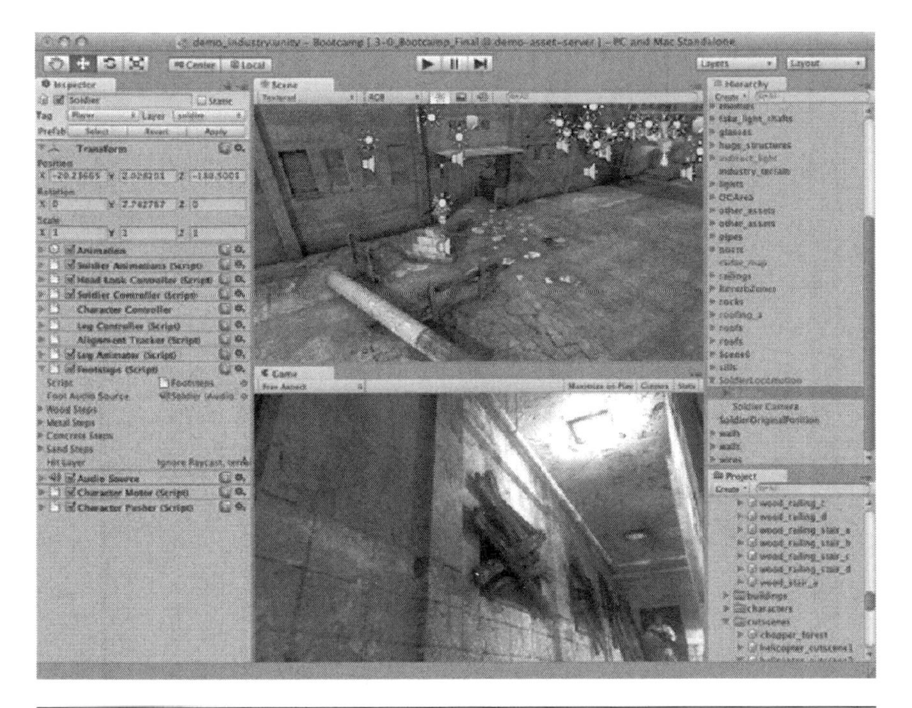

Figure 9.23 An authoring system for immersive and natural UI-based contents (Unity3D). (From Unity3d, Unity Korea, http://www.unity3d.com.)

Critical to such a future vision will be the VR/immersive/natural UI–based content-production pipeline, starting with content-authoring tools. Such tools with capabilities beyond just game development or multimedia editing are already starting to appear (Figure 9.23).

9.5 Mixed and Augmented Reality

Mixed and augmented reality is yet another interaction medium receiving lots of hype these days. Mixed and augmented reality refers to the medium in which the representations of the real and virtual are mixed in some proportion (the term *Virtuality or Mixed Reality Continuum* was coined accordingly [9]; see Figure 9.24). For example, content with mostly real objects and only a small portion of the virtual is called *augmented reality*, while the reverse is called the *augmented virtuality*. MAR requires a few core technologies, namely object recognition and tracking (Sections 9.1.2 and 9.1.3). This is required to spatially register the augmentation right next to the object targeted for augmentation. A looser form of MAR simply

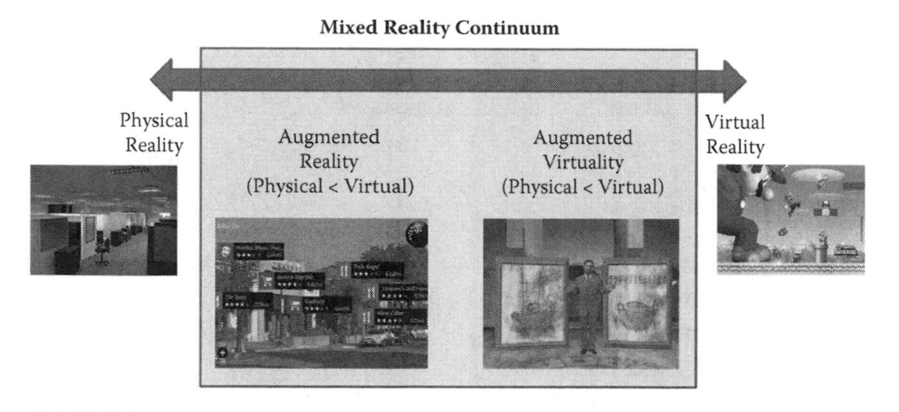

Figure 9.24 Mixed reality/virtuality continuum [10]. A spectrum is formed according to the relative proportion of the real and virtual representations in the content. At the extremes of the continuum, there is the completely real environment and the purely virtual environment.

augments the information anywhere on the screen. A Google Glass type of application is such an example, where information is projected on the see-through glass at a fixed position (top right corner of the visual field). MAR can improve the usability and UX in interacting with everyday objects because the associated information resides and is displayed at the same location, with the possibility of instant recognition and access.

9.6 Others

We have briefly looked at several promising technologies and future trends for HCI (in this very subjective view by the author). There are certainly others (which have actually been touted as interfaces of the next generation), which I have not cared to advocate due to various perspectives of my own. I will briefly go over them here before wrapping up this book.

- *Wearable computing and interaction*: The smartphone, while almost undetachable from many users, is not a true form of wearable computer. The concept of a wearable computer started from the idea of embedding computers and interaction devices into clothes and things we wear (e.g., hats, belts, shoes, glasses). This integration of "wears" and computing devices has not advanced as much as expected during the last decades in terms of both technology and usability. Even the

Google Glass concept is facing practical problems such as its weight, power, and privacy issues. It is still questionable whether computer elements need to be interwoven into our "wears" (except for very special applications).

- *Interaction based on physiological signals*: Much research has been conducted in ways to take advantage of our physiological signals such as brain waves, EMG (electromyography), ECG (electrocardiography), and EEG (electroencephalography). It seems very difficult to extract human intention in a useful and major way for HCI from these raw signals. This line of research will probably focus on the HCI for disabled people.
- *Eye/gaze tracking and interaction*: HCI is deeply connected with the line of sight. When interacting, we mostly tend to look at the target interaction object. Tracking of the line of sight is often done by tracking the head direction, rather than the eyeballs themselves. In many cases, it is safe to assume that the front head direction is the direction the eyes are looking. There are not too many applications in which the exact eyeball/gaze direction is so important (except maybe for gaze analysis).
- *Facial/emotion based input*: Affective interfaces based on aesthetic look and feel and on more humane output feedback may be important and emerging techniques for improving UX. However, as an input method, it seems we have a long way to go. Input based on user emotion (e.g., facial expression, tone of voice, particular gestures) is very difficult even for humans themselves, and thus would be very difficult to be used as a robust means of interaction.
- *Finger-based interaction*: As explained in Section 9.1.2, finger-based interaction has been pursued through the use of gloves. Depth-based sensing has recently allowed finger tracking and interaction without the inconvenience of having to wear a glove. Again, not too many applications can be found where finger-based interaction can be applied in a natural way. Contrived finger gestures can be used, but they generally incur low usability.
- *3-D/stereoscopic GUI*: Interacting by manipulating 3-D GUIs (in stereo) has been depicted in many science fiction movies. However, there are not many computer tasks that require

precise 3-D motions. Most system commands are easier with voice or the familiar 2-D cursor control.

- *Context-based interaction*: Similar to the case with the emotion-based input, inferring "context" in hopes of adapting to the operational situation at hand or of personalizing the interface to the user is very difficult. The true user intent is not always clearly manifested explicitly and capturable/interpretable by the sensors and AI.

9.7 Summary

The utility of software and digital content will increasingly depend on HCI capabilities and less on the core functionalities of conventional computer hardware. The HCI issue is becoming more challenging as the number of computing platforms proliferates to accommodate the evolving usage situations (e.g., home, office, mobile, sales, vehicles, military, etc.). Design of the HCI interface will continue to play a significant role as the influence of the standard desktop platform declines. Better design of HCI interfaces will give everyone better access to available services, intelligence, and knowledge. We will have power at our fingertips.

References

1. IBM. 2014. IBM Watson. http://www-03.ibm.com/innovation/us/watson/.
2. Wikipedia. 2014. Google voice search. http://en.wikipedia.org/wiki/Google_Voice_Search.
3. Apple. 2013. iOS 7, Siri. http://www.apple.com/kr/ios/siri/.
4. AT&T Labs Research. 2014. AT&T WATSON(SM) Speech Technologies. http://www.research.att.com/projects/WATSON/?fbid=j6ZWSYBnql4.
5. Electronic Arts. 2014. Black and White 2: Battle of the gods. http://www.ea.com/black-and-white-2-battle-of-the-gods.
6 Engadget. 2013. PrimeSense demonstrates Capri 3D Sensor on Nexus 10. http://www.engadget.com/2013/05/15/primesense-demonstrates-capri-3d-sensor.
7. Leap Motion. 2014. Leap Motion controller. http://www.leapmotion.com.
8. Bolt, Richard A. 1980. Put-that-there: Voice and gesture at the graphics interface. In *Proceedings of ACM SIGGRAPH*, 262–70. New York: Association for Computing Machinery.

9. Milgram, Paul, H. Takemura, A. Utsumi, and F. Kishino. 1995. Augmented reality: A class of displays on the reality–virtuality continuum. In *Proceedings of the International Society for Optics and Photonics for Industrial Applications*, 283–92. Bellingham, WA: SPIE.
10. Wikipedia. 2014. Reality–virtuality continuum. http://en.wikipedia.org/wiki/Reality%E2%80%93virtuality__continuum.

Index

N

O

P

Q

R